TSERING STOBDAN

Agriculture in Ladakh

A Step Towards Sustainable Mountain Development

THRIVING AGAINST THE ODDS:
FARMING IN THE COLD DESERT

ISBN : 978-93-95266-76-5

Published by: Beeja House
Printed By: Repro India Pvt. Ltd.

First Printing Edition 2023
Author Email: tstobdan@gmail.com

Cover photo by: Tsering Stobdan

This book is dedicated to the farmers – the backbone of Ladakh's agriculture

CONTENTS

PREFACE

"What do people eat in Ladakh?"; "Why apricots of Ladakh are so sweet?"; "Are the crops grown in Ladakh organic?"; "What research topic shall I choose for my Ph.D. thesis?". People generally ask these questions when they know I am an agriculture scientist at the Defence Institute of High Altitude Research, Leh Ladakh. My answers to these questions end up in a few sentences, and on many occasions, the discussion continues for hours. In the middle of the discussions or after a lengthy conversation, I feel the need to put the facts and my ideas on paper to effectively answer their questions the next time.

I had the opportunity to write several reports on topics related to organic farming, seabuckthorn, apricots, greenhouses, and insect pests. However, the circulation of these reports remains restricted to a few individuals and agencies. And I have always wished for wider dissemination of the core message espoused in these reports, targeting a larger audience so that they, too, can contribute to achieving the goals of sustainable agriculture in Ladakh.

Subsistence farming is traditionally practiced in Ladakh, and the region was self-sufficient until the 1960s. However, it has been widely accepted that the current agricultural system cannot help farmers sustain themselves on small and marginal lands. Agrarian reforms are unavoidable to cope with the changing scenario. The farm products of Ladakh are considered at par with that of the rest of the country, which makes the product a 'commodity' rather than a 'premium produce.' Something needs to be done so that Ladakh's agricultural products are considered special and unique, bringing them extra value.

During my two decades of working at the grassroots level and with people's representatives, administrators, and extension workers, the teamwork which the people of Ladakh have shown are lessons of relevance far beyond Ladakh itself. Many of the technologies we developed at our research farms have been successfully replicated at the farmers' field in the shortest possible time, all possible due to teamwork. I have shared some of my personal experiences in this book as well.

The agricultural sector needs accurate data to help the individuals and agencies involved in the agricultural chain understand every detail during the process. I want to thank the Chief Agriculture Officer, Chief Horticulture Officer, and Deputy Registrar Cooperative of Leh & Kargil districts for sharing updated crop production data, land use patterns, and input distributions.

In my attempt to write about the century-old traditional farming system, I took the help of knowledgeable persons with vast grassroots experience. I sincerely thank Mr. Viliat Ali, Naib Tehsildar, for sharing his wisdom. Thanks to Mr. Tsering Phuntsog, Mr. Stanzin Kunkhen, and Dr. Naveen Bisht for their help in editing a few chapters. I am thankful to the renowned filmmaker Mr. Stanzin Dorjai Gya for the permission to use some of his photographs in this book.

A major part of this book is based on the research carried out at the DIHAR for over five decades. I am thankful to the Director and colleagues at DIHAR for their constant, unwavering support. Thanks to my Ph.D. scholars who generated new knowledge on various agriculture-related topics. Thank you, Geetika and her team. I appreciate your mentoring and showing me how to write this book step-by-step.

This book would not have been possible without the love and support of my family- Dekyi, Chosdan, and Nawang!

Chapter-1

Ladakh: Topography, Climate, and Soil

A village landscape with mustard and barley fields

Ladakh is a high-altitude desert crisscrossed by a massive mountain range. The name "Ladakh" is probably derived from the Tibetan *la-dags* meaning 'the land of mountain passes'. Under the British Indian Empire, Ladakh was considered an insignificant, remote, deserted rural area in trans-Himalaya (Pelliciardi, 2010). Ladakh became a Union Territory on the 31st of October 2019.

Situated in the trans-Himalaya, between Karakoram (North), the Himalayan ranges (South and West), and the Tibetan Plateau (East), it is geographically classified as a high "cold desert" (Negi, 2002). Ladakh comprises two districts, namely Leh and Kargil. The altitude of the Leh district ranged from 2900 to 5900 m, while that of the Kargil district is 2400 to 4300 m above mean sea level. The geographical area of the Leh district is 45110 km^2, while that of Kargil is 14036 km^2. There are 113 revenue villages in Leh and 130 villages in Kargil district.

As per the 2011 census, the total population is 2,74,289, equally divided into Leh (1,33,487) and Kargil (1,40,802) districts. The population of Leh is predominantly Buddhist, while in Kargil, most of the population follows Shia Islam. Despite the vast geographical area, 59.6 percent of the households in the Leh district and 82.5 percent in the Kargil district have less than one hectare of cultivable land. The size of a village is generally determined by the amount of meltwater supply from glaciers or permanent snowfields in the upper catchments. Because of the cold desert condition, only 0.4 percent of the total geographical area is under cultivation, and the area under forest cover is a meager 0.236 percent.

The seasons dictate life in Ladakh more than in any other inhabited place. Scorched by the sun in summer, the entire region freezes solid in winter when the temperature drops to as low as minus forty degree Celsius (Norberg-Hodge, 1991). Leh town's hottest month is July (25.8°C), and January is the coldest (-13.0°C). The agricultural cycle begins between February and June, depending on the altitude. The maximum and minimum temperature during the cropping season (May-September) is 22.3°C and 9.4°C, respectively. Generally, April marks the beginning of the cropping season, while mid-September marks the end of the harvest. Early sowing or transplanting often results in high mortality of seedlings due to spring frost, and late harvesting after mid-September often results in freezing injury.

Ladakh lies in the rain shadow of the Himalayas and thus receives little rainfall. The average annual precipitation is less than 200 mm, of which more than 70% is snowfall. The region experienced intense sunlight and high UV-B radiation. The average light intensity at noon during cropping season is 150-kilo lux. The UV-B, expressed in terms of minimum erythemal dose per hour (MED/Hr), is significantly higher in magnitude as well as in time duration in Ladakh compared to Antarctica and Delhi. On a hot summer day, the estimated UV-B at local noon time in Leh is around 4.0 MED/Hr. In comparison, the UV-B in Delhi is 2.0 MED/Hr. Therefore, UV-B radiation in Leh is double compared to Delhi (Singh et al., 2005).

By and large, the soils are shallow to very shallow in depth. The soils usually consist of coarse materials and are often called Skeletal soils with admixtures of distinguished rock fragments, colluvial materials, and screes. The soil is alkaline in reaction, with pH ranging from 8.3 to 8.6, and low in organic carbon (0.16 to 0.58%) (Gupta and Arora, 2017). The agricultural land is medium in organic carbon and potassium and low in phosphorus. It is deficient in micronutrients such as zinc, manganese, and iron (Dwivedi et al., 2005).

REFERENCES

Dwivedi SK, Sharma VK (2005). Status of available nutrients in the soil of cold arid region of Ladakh. *J Indian Soc Soil Sci.*, 53: 421–423

Gupta RD, Arora S (2017). Characteristics of the soils of Ladakh region of Jammu and Kashmir. *J Soil Water Conserve.,* 16(3): 260–266

Negi SS (2002). *Cold desert of India.* Indus Publishing Company, New Delhi

Norberg-Hodge H (1991). *Ancient futures: Lessons from Ladakh for a globalizing world.* Sierra Club Books, San Francisco

Pelliciardi V (2010). Sustainability perspectives of development in Leh district (Ladakh, Indian Trans-Himalaya): an assessment. A thesis submitted to the University of Rome for the award of Doctor of Philosophy

Singh R, Nath S, Tanwar R, Singh S (2005). Study of erythemal dose variation and exposure time for different UV-B dose levels at Indian mainland and Antarctica. *Proc. URSI*, New Delhi

Chapter-2

Agriculture in Ladakh: An Overview

Women and children carrying Labtsey, harvested barley, for thrashing

Today Ladakh is known as a tourist destination, but the region is primarily an agrarian community with roughly 70 percent of the population directly or indirectly dependent on agriculture. For many centuries the people of Ladakh have led a self-reliant existence mainly based upon subsistence agriculture, pastoralism, and caravan trade. However, with population growth from 1,05,292 in 1971 to 2,74,289 in 2011, the region is no longer self-sufficient in food production. The existing cropped area is sufficient to meet the food requirement of roughly

60,000 people, and thus there is a considerable gap between demand and supply.

Traditional farming is integral to an intangible heritage passed from generation to generation. It matters beyond its economic contribution. It is the root and foundation of Ladakhi society, which blooms and prospers in hostile and barren terrain. The traditional close ties between agriculture and cultural development can still be seen and felt in Ladakh. Small and fragmented holdings characterize the agriculture system in Ladakh. Despite the vast geographical area, 59.6 percent in the Leh district and 82.5 percent of the households in the Kargil district have less than one hectare of cultivable land. Single-cropping is dominant, as double-cropping is possible only in a limited area falling below 3000 m of altitude. Agriculture production is based entirely on irrigation. Glaciers are the primary water source for irrigation, and the rivers that flow in the region remain underutilized for agricultural purposes. The region remains cut off for over five months in a year due to heavy snowfall. Availability of locally grown fresh farm produce is restricted to summer months; therefore, there are seasonal differences in dietary food intake. Due to its unique climatic conditions, there is a glut of farm produce in August and September, while there is a total absence in the availability of fresh fruits and vegetables during the rest of the year.

Domestic food production is unable to meet the demand of the increasing population. Leh district gets approximately 73 percent of its food grain outside the region. The vegetable import dependency is roughly 67 percent, while that of fruit is about 85 percent. Therefore, self-sufficiency in food is an important issue for the region. Meeting the demand of the region through local production is a difficult task. Importing goods to Ladakh necessitates shipping by trucks across the Himalayas, with passes as high as 5300 m, from Manali or Srinagar.

The total cropped area in Ladakh is 23,612 hectares (Leh: 10,358 ha; Kargil: 13,254 ha); thus, only 0.4 percent of the total geographical area is under cultivation. Similarly, the area under forest cover is a meager 0.236 percent. One frequently asked question is: how much additional land could be used for crop production? Data suggested that an additional 35,424 hectares (Leh: 26,836 hectares; Kargil: 8,588 hectares) will be available for cultivation in the next three decades, which is more than the existing cropped area in the region.

2.1 Major crops

Cereal crops: Wheat and barley are the traditional crops of Ladakh and continue to be the staple food of the people. Cereals are grown for twin purposes of food and fodder. Of the 23,612 hectares under cultivation, barley (10,082 hectares) and wheat (4,512 hectares) occupy 61.8 percent of the total cropped areas. The region produced 17,056 tonnes of barley and 7,020 tonnes of wheat in 2022. However, since wheat is readily available from the Public Distribution System (PDS), farmers are gradually becoming less interested in cultivating wheat crops.

Vegetables: In the late 1960s, only few types of vegetables were grown in the region. Recently, the feasibility of growing 101 types of vegetables has been demonstrated. Farmers in Ladakh are growing 23 types of vegetables commercially and supplying them to the Army in the region through the Farmers' Cooperative Marketing Society. The Ladakh region produces off-season vegetables such as broccoli, cabbage, cauliflower, and peas. However, due to poor market linkage, large-scale production of vegetables as off-season crops has yet to gain momentum in the region. The region produced 17,572 tonnes of vegetables in 2022 (Leh: 9,071.2 tonnes; Kargil: 8,500.8 tonnes) from 1,163 hectares.

Fruits: The climatic condition is highly conducive for producing quality apricot and apple. Historically, premium-quality dried apricots have been used as one of the primary trading commodities with neighboring countries. Recently, *Raktsey Karpo*, apricots with white seed stone, have been identified as a unique genetic resource of Ladakh and the world's sweetest apricots. There is a potential for large-scale production and marketing of quality organic apricot and apple from Ladakh.

Apricots and apples are the two major fruit crops of the region and are prized for their quality. Of the 23,612 hectares under cultivation, apricot (2,612.6 hectares) and apple (946.6 hectares) occupy 15.1 percent of the total cropped areas. The region produced 15,864 tonnes of fresh apricot (Leh: 5,059 tonnes; Kargil: 10,805 tonnes) in 2022, making it the largest producer of apricot in the country. The total production of fresh apples is 5,191 tonnes (Leh: 3,221 tonnes; Kargil: 1,970 tonnes).

Fruit crops such as grapes, peach, plum, mulberry, and walnut are also being grown in the region. However, production volume is low, and it is hard to find such products in the local market. Melons have recently been introduced in the region as a cash crop. Seabuckthorn (*Hippophae rhamnoides* L.) grows naturally in Ladakh without much human interference. Currently, the demand for Seabuckthorn in the region exceeds the supply. Seabuckthorn has recently been included as a horticultural activity under the Mission for Integrated Development of Horticulture (MIDH) scheme of the Ministry of Agriculture and Farmers Welfare, Government of India.

Pulses: Pulses are grown on 1,168 hectares in Ladakh. The region produced 886 tonnes of pulses in 2022 (Leh: 135.8 tonnes; Kargil: 750.8 tonnes). The Agriculture Department is promoting cultivating pulses in the region, and therefore, there is an increase in the area under pulses. *Rajma* (kidney beans) grown in the region is of high quality and is found to be with a delectable taste, and is cooked quickly. It is a low-risk crop and has the potential for marketing as a premium organic produce of Ladakh.

Oil seeds and buckwheat: The cultivation of oil seeds and buckwheat is restricted to small areas only. The region produced 115 tonnes of oil seeds (Leh: 106 tonnes; Kargil: 25.9 tonnes) and 486.4 tonnes of buckwheat (Leh: 384.8 tonnes; Kargil: 101.6 tonnes).

Fodder: Fodder is the backbone of Ladakh's agriculture and animal husbandry industry. Cultivated fodder occupies 4,915 hectares of cropped areas. The region produced 50,737 tonnes of fodder in 2022 (Leh: 34,906 tonnes, Kargil: 15,831 tonnes). The region is known for high-quality alfalfa, a rich protein source. Dried alfalfa is traditionally used as the primary source of fodder during winter months.

2.2 Protected cultivation

Given the necessity of growing vegetables in winter, many passive solar greenhouses have been established in Ladakh since the 1980s. Some commonly known greenhouses are Ladakhi polyhouse, trench, polycarbonate, FRP, and tunnel. The two widely used greenhouse designs for the Ladakh region are the traditional Ladakhi polyhouse and the recently developed Ladakh Greenhouse. The traditional Ladakhi polyhouse has been in use since the 1980s, while the Ladakh Greenhouse is being established on farmers' fields from 2020 onwards. The increased use of the greenhouse has not only improved the dietary intake of vegetables during the winter months but also opened a new economic opportunity for the sale of early-season vegetables by the local farmers.

2.3 Livestock

Farmers primarily practice mixed farming, where livestock is reared as an integral part of the system for food and manure. There are 4,02,329 livestock in the Leh district and 3,82,250 in the Kargil district.

2.4 Major insect pests and diseases

The incidence of insect pests and diseases is low due to low temperatures and relative humidity. However, the incidence of codling moths, leaf-curling aphids, yellow-tail moths (*Euproctis similis*), and gummosis are major problems of fruit trees. Loose smut is a major disease of cereals in the region, while in vegetables, the incidence of cutworms, onion maggots, aphids, and cabbage butterflies are major insect pests.

2.5 Fertilizer and pesticide use

The application of fertilizer in Ladakh is relatively low. While the national average fertilizer consumption during 2021–22 was 146.7, it was 40.9 kg per hectare in Ladakh (Leh: 30.8 kg per hectare; Kargil: 48.7 kg per hectare). There has been a considerable reduction in overall fertilizer use in the Leh district from 732.0 tonnes in 2012 to 320 tonnes in 2022. Similarly, fertilizer use in the Kargil district has reduced from 874 tonnes in 2012 to 646 tonnes in 2022. The reduction is primarily due to the organic movements in the region and the adoption of the Mission Organic Development Initiative of Ladakh.

2.6 Water resources

Glacier-fed and snow-fed gravity-controlled irrigation systems are predominant in Ladakh. Adequate meltwater supply from glaciers or permanent snowfields in the upper catchments or, where local topographical conditions permit, direct abstraction from the mainstream is a fundamental prerequisite for crop production in Ladakh. Glaciers provide most of the runoff in summer, with a maximum regularly reached in the afternoon. In such water surplus situations, the evening and nighttime runoff is usually stored in reservoirs (*Zing*) and diverted to the fields in the morning when the natural runoff is scarce (Nüsser et al., 2012). Severe water scarcity occurs during the sowing period in April and May due to snow-deficient winters or delayed snowmelt. A sufficient and reliable meltwater supply from high-altitude glaciers becomes available from June to October.

2.7 Key challenges

Ladakh currently faces a multitude of challenges related to agriculture: increasing affluent population, rural-urban migration, shrinking agricultural land, shrinking agro-biodiversity, increased dependency on imported food, scarcity during winter, seasonal availability of fruits and vegetables, post-harvest losses, poor market access, threatened traditional farming cultures and climate change. The majority of people in Ladakh live in rural areas. However, the region is experiencing rapid population growth and declining per capita farm sizes. The population of Ladakh has increased from 1,05,292 in 1971 to 2,74,289 in 2011. The urban population of Ladakh has seen a rise from 7.5 percent in 1971 to 22.6 percent in 2011. Moreover, Ladakh's floating population is increasing rapidly. The value of land has increased to the point where the return from agriculture/horticulture production cannot compete with other developmental activities, especially in the areas coming under urbanization. The challenges are multidimensional and require coordinated and targeted initiatives that can contribute to achieving sustainable agriculture development.

2.8 The flagship program

In 2019, the Ladakh Autonomous Hill Development Council (LAHDC), Leh, took its first step to embrace organic farming. A Special General Council meeting of the LAHDC Leh was held on 9 March 2019, and the Council unanimously declared it to become a fully organic district by 2025. The Administrative of the Union Territory of Ladakh launched a Rs 500-crore organic farming project named 'Mission Organic Development Initiative of Ladakh' on 4 July 2020 for Leh and Kargil districts. The project also aims to realize the goals of a carbon-neutral Ladakh. It seeks to transform agriculture in Ladakh into a sustainable, remunerative, respectable occupation and to enable the farmers to reap the benefits of dynamic market opportunities.

REFERENCES

Angmo S, Dolkar D, Dolkar P, Kumar B, Stobdan T (2019). Growing watermelon in high altitude trans-Himalayan Ladakh. *Natl Acad Sci Lett.*, 42: 379–382

Ladakh Autonomous Hill Development Council (LAHDC), Leh Ladakh (2019) Ladakh organic policy document 'Mission for Organic Development Initiative of Ladakh: Policy, Strategy, and Action Plan'

Nüsser M, Schmidt S, Dame J (2012) Irrigation and Development in the Upper Indus Basin. *Mt Res Dev.*, 32: 51–61

Peaty, D. (2009). Community-Based Tourism in the Indian Himalaya: Homestays and Lodges. *J Ritsumeikan Soc Sci Humanit.*, 2: 25–44

Barley harvesting in Gya village

Chapter-3

The Traditional Farming System in Ladakh

Ploughing field with a pair Dzo, a male hybrid of yak and cow

The century-old traditional farming system in Ladakh is an established agricultural practice and indigenous knowledge handed orally from parents to children over the centuries. This system is still in use and coexists alongside modern farming systems. Despite the rough topography and the harsh climatic condition, the people of Ladakh have learned to live with nature. The farming system was refined over the years, and the unproductive soil was turned into productive fields. The adventurous Moorcroft stayed for two years in Ladakh, from

September 1820 to September 1822, stated in his travelogue: "*Notwithstanding these unpromising conditions the harvests of Ladakh are by no means unproductive, and they present also the peculiarity of yielding equally abundant crops year after year from the same land, without its ever being suffered to lie fallow, and without any attempt being made to cultivate a succession or alternation of produce*". He also stated in his travelogue: "*In no other country have I seen an equal surface in barley as regularly covered with plant, and never plants with better heads*" (Moorcroft and Trebeck, 1837).

Most of the farmers are self-supporting, living in small settlements scattered throughout the vast geographical area. Farmers carry out a collective workforce-sharing system called *Langde*, through which all agricultural activities are completed without hiring laborers. The agriculture fields are small in size due to the mountainous terrain. This also necessitated the construction of terraces which are supported by a stone breastwork. All cultivable land is called *Zhing*. Fertile land that gives high crop yield is called *Zhing-zang*, while those on rocky land are called *Ri-zhing*. The prime agriculture field of a household is called the *Ma-zhing*. Medium fertile fields are called *Bar-zhing*, while the least fertile field is known as *Tha-zhing*. Some fields are cultivated in the alternate year due to low fertility and are called *Sa-tsik*. Those fields where only vegetables are cultivated are known as *Tsas*. Fields where fodder is cultivated, are called *Olthang*.

Agriculture is entirely dependent on irrigation. All cultivable land lies along the course of small streams and river banks that originate from the glaciers. The waters from the smaller stream are collected in ponds known as *Zing,* from where the water is systematically channelized to irrigate all the fields downstream. Cunningham mentioned in his travelogue: "*These canals, which are conducted several hundred feet above the villages, are mostly built up with a retaining wall, and puddled with clay to hold the water. In a few places, the rock itself was excavated to form a passage for the water, but in other places, where the hill was too precipitous, or the rock was too hard, the water was passed along hollow poplar and willow trunks, which were supported by uprights standing on ledges of the rock, or on huge pegs driven into its crevices*" (Cunningham, 1853).

Centuries-old water management systems ensure that the water is carefully distributed so everyone can have their fair share of the precious resource. Each year, a group of people among the villagers is appointed or elected as *Churpon* (Waterlord), whose main job is to manage the proper distribution of the village water.

The land is measured in terms of *Khál* or *Nyima.* One *khál* is the size of the field that requires one *khál* of seed, which is approximately 12 kg. Area wise, a *khál* of land is roughly equal to one-sixteenth of an acre. Agriculture land is also measured by *Nyima,* meaning a day-long or a field size that requires a day-long (8 AM to 6 PM) ploughing by a pair *Dzo,* a male hybrid of yak and cow. So, one *Nyima* field is roughly equal to three-fourth of an acre. Women are the principal laborers in the field as the female population is higher than the male population. The main crops cultivated in Ladakh are wheat, barley, and buckwheat. The generic name for wheat is *Tro*, while that of barley is *Nas*. Buckwheat is referred to as *Bro*.

Barley is grown in villages at high as well as low elevations. It grows even at 15,000 feet or above in the Changthang region. However, wheat is grown only at low elevated areas below 12,000 feet. Barley crop requires sixty to hundred days from sowing to harvesting, while wheat needs approximately one hundred and forty days. The sowing of wheat is done fifteen to twenty days before barley and is harvested two to three weeks after reaping the barley. Buckwheat is grown at lower altitudes as a second crop after the harvest of barley. It requires six weeks to two months from sowing to harvesting.

The wheat is of four types: *Tro Chen* (red wheat), *Tro Karmo* (early wheat), *Tro Surutze*, and *Harosa* wheat. *Tro Karmo* is the most productive and yields the finest flour. The barley is of two broad types –husked and naked. Husked barley is called *Nas Swa*, while the huskless or naked barley is called *Sherokh* or *Yangma*. When the *Sherokh* is quite ripe, the grain shrinks and remains loosely attached. With a slight shake, the grain gets disposed of. If the crop is left standing after it is fully ripe, the grain is more disposed to shed than common barley. Therefore, *Sherokh* is generally harvested before it is perfectly matured. *Sherokh* barley is preferred in Ladakh to the common, or husked barley, for all uses, but especially for preparing the fermented liquor called *Chhang* (Moorcroft and Trebeck, 1837).

Six types of *Sherokh* are grown in Ladakh: *Chu Nas* (slow or late barley), *Giok Nas* (early barley), *Nas Yan Karmo* (early), *Nak Nas* (black barley), *Tughzur Nas* (six-sided barley), and *Mentok Nas* (flower barley). *Nas Yan Karmo*, sometimes denominated as *Sarmo*, is cultivated everywhere. *Giok Nas* is grown in places having short summers. *Nak Nas* is the hardiest of all types of *Sherokh*, and grows at the most extreme altitudes (Moorcroft and Trebeck, 1837).

Nas Swa is generally grown in warmer places such as Thiksey and Chuchot villages near Leh. Though farmers prefer to grow *Sherokh*, growing *Sherokh* continuously for three years turns them into *Nas Swa*. Therefore, frequent seed replacement is required in such places to get *Sherokh* variety.

The agricultural calendar is based on the astrological facts of *Lotho* (Tibetan almanac) (Angchok and Dubey, 2006). Astrologers are consulted to begin the agriculture activity, referred to as *Saka*, on an auspicious day. The spirits of the earth and water – the *sadak* and the *lhu* – are pacified before ploughing the field (Norberg-Hodge, 1991).

The farming practices differ slightly from place to place depending on the altitude and availability of resources. Farming systems traditionally practiced in villages located at a higher elevation, above 12,000 feet, such as Nang, Stakmo, Saboo, and Gangles around Leh town, are summarized below. In these villages, only barley is traditionally grown due to short growing season.

Ploughing (*Zhing-log*)

The agriculture cycle begins between February and June, depending on the altitude. The springtime ploughing called *Spid-log* is done in late March to early April in villages located at higher elevations. In lower areas, the ploughing is carried out in October, known as *Ston-log*, roughly forty days after the harvest, in place of *Spid-log*. Ploughing is done with a pair of *Dzo*. Six to eight inches deep furrows are made with the help of a wooden plough known as *Showl* that has a metal share. The plough is a simple device that requires no major maintenance except the iron share, which requires frequent sharpening. In the Changthang region, the ploughing is performed by a pair of yaks, while horse-drawn ploughing is also carried out in areas such as Turtuk and Nyoma.

Manuring (*Lúd*)

After the land has been once ploughed, animal-based manure, called *lúd* is applied in the fields. Composting night soil from traditional dry toilets, known as *Chak-lúd*, is also used as manure. Manure obtained exclusively from goat is known as *Ra-lúd* while that of cow is called the *Ba-lúd*. All kinds of animal dung and ash are used as manure. The manures are transported to the field on the back of donkeys.

Irrigation (*Tha-chus*)

After the land has been ploughed, the field is soaked with irrigation before sowing in the spring. This irrigation is called *Tha-chus*. If the area receives sufficient snowfall during sowing, a *Tha-chus* is not required. The sowing process called *Kha-ser* is carried out on a snow-moistened field.

Second ploughing (*Zhing-gnos*) and sowing (*Nyoches*)

Roughly about eight to ten days after *Tha-chus*, the manure which is brought to the field is scattered uniformly a day before sowing. Sowing, which is known as *Nyoches*, is done in the furrows made by ploughing, which is carried out for the second time. This process is called *Zhing-nyos*. Sowing in the furrow is called *Roltab*, while in some parts of Ladakh, such as the Sham and Turtuk area, a broadcast method called *Gyastor* is also practiced. The seed required for *Roltab* is roughly twice that of *Gyastor*. After sowing, the field is levelled with a rake-like device called *Bhad*.

Compartmentation (*Shau-jang-ches*)

When the shoots grow to about an inch long, the field is divided into compartments to facilitate uniform irrigation. This process is known as *Shau-jang-ches*.

Weeding (*Yurma-yurches*)

Yurma-yurches or the weeding is also carried out after about three weeks. However, the majority of the farmers do not do the weeding.

The first irrigation (*Dol-chu*)

Dol-chu, the first irrigation, is carried out when the shoots reach about two inches from the ground. It is generally done one month after sowing, i.e., mid-June. Considerable skill and care are required for the first irrigation, as excess watering can damage the tender crop.

The second irrigation (*sRhak-chu*)

Ten days after the *Dol-chu* a shallow irrigation known as *sRrak-chu* is carried out for the second time.

The third irrigation (*Non-chu*)

The irrigation for the third time is done a week after the second irrigation. This is known as *Non-chu*, and hereafter, the irrigation is done at weekly intervals.

The last irrigation (*Dro-chu*)

Dro-chu, the last irrigation, is done a week before the harvest. In general, barley crops are roughly irrigated ten times.

Harvesting (*gNab-sa*)

gNab-sa, or the harvesting of barley, is carried out in September. The barley crop is considered ideal if it is harvested after 100 days. A 100-day barley crop is considered best in terms of taste and yield.

Harvesting is done with the help of *Zora* (sickle). The stem is cut close to the ground leaving the root in the ground. However, in some places in Kargil and Sham in lower Ladakh, the crop is pulled along with the root. The harvested crop, known as *Labtsey*, is kept on the field upside down for about two days, known as *Porotsey*. There is a tradition that some finest ears of barley are brought home and fastened to the pillars in the house as a sign of a good harvest.

Thrashing (*Khoyus*)

The harvested crop is brought to a place where thrashing is to be carried out. A circular place of about thirty feet in diameter is made, which is known as *Youltak*. The harvested crop is brought near the *Youltak*, placed in heaps called *Choke*, and left undisturbed for about two weeks for further drying. The thrashing is carried out with the help of animals. On the day of *Khoyus* (thrashing), seven to eleven heaps of *Chokes* are placed on *Youltak*, and a number of animals tied to a rope attached to a central pole are made to trample the crop as they walk. A combination of seven to eleven animals, such as cow, *Dzo*, horse, and donkey, is used for thrashing. The cow is placed next to the pole, the donkey and *Dzo* in the middle, while the horse who runs the most is placed furthest to the circle. Only two thrashing cycles are done in a day to prevent overexertion of the animals.

Winnowing (*Ongs charches*)

Winnowing is done in a perfect rhythm by two people facing each other carrying wooden forks, called the *Zar*. The crop is scooped into the air so that the wind blows away the chaff, and the heavier part containing the grain (*Ongs*) falls to the ground. The *ongs* is then sieved on *tholma* to separate the grain. The grain is packed in sacks and those with ears known as *Achik* still intact are kept separated and subjected to thrashing again after about ten to fifteen days.

The final harvest is measured in terms of *khál*. The average return varies from five to ten-fold depending on soil fertility and altitude. In general, six-fold harvest is obtained in barley and eight-fold in wheat. Significantly higher yield is obtained in places that employ the broadcast sowing method.

REFERENCES

Angchok D, Dubey VK (2006). Traditional method of rainfall prediction through Almanacs in Ladakh. *Ind J Trad Know.*, 5: 145–150

Cunningham A (1853). *Ladakh: Physical, statistical, and historical with notices of the surrounding countries.* Edition 1997. Gulshan Publisher, Gow Kadal, Srinagar, India

Moorcroft W, Trebeck G (1837). *Travels in the Himalayan provinces of Hindustan and the Panjab in Ladakh and Kashmir; in Peshawar, Kabul, Kundiz, and Bokhara.* Gyan Publishing House, New Delhi-110002

Norberg-Hodge H (1991). *Ancient futures: Lessons from Ladakh for a globalizing world.* Sierra Club Book, San Francisco.

Transporting livestock manure (Lúd) to the field on the back of donkey

Chapter-4

Vegetable Production Scenario in Ladakh

Vegetables are an indispensable part of a healthy diet, and variety is as important as quantity. Since each vegetable contains a unique combination of phytonutriceuticals, a great variety of vegetables should be consumed. Accordingly, diets with a high quantity of vegetables are widely recommended. Most countries have dietary recommendations for vegetables. A healthy diet needs to include nearly 500 g of vegetables and fruits as per the recommendation of the National Institute of Nutrition. However, more than 80 percent of Indians do not meet this recommendation. The recommended quantity of vegetables for an Indian soldier in high altitude is 140 g potato, 60 g onion, and 170 g fresh vegetables per head per day (Mishra et al., 2010).

In Ladakh, the availability of fresh vegetables remains restricted to summer months. Long, harsh winters reduce the cropping season to just four to five months a year, and the region remains cut off for over five months a year due to heavy snowfall. Therefore, there are seasonal differences in dietary intake of food. Fresh vegetable consumption decreased significantly during winter, resulting in micronutrient deficiencies, a phenomenon that has been described as 'hidden hunger' (Dame and Nüsser, 2011). A survey conducted in the year 2008 found that only 19.4 percent of the people in Ladakh are well nourished, and 35 percent of people suffer from malnutrition-related diseases. The most prevalent diseases are anemia (8.7 percent), followed by night blindness (7.3 percent), scurvy (6.7 percent), beriberi (6 percent), pellagra (3.7 percent), and rickets (3 percent of the surveyed subjects) (Dar and Rather, 2014).

Self-sufficiency in fresh vegetables is an important issue for the region. Importing vegetables to Ladakh during summer necessitates the shipping of vegetables by truck across the Himalayas, with passes as high as 5300 m, covering the distance of Manali to Leh (430 km) or Srinagar to Leh (420 km) (Pelliciardi, 2013). During winter, commercial aircraft bring approximately 150–200 tonnes of fresh vegetables, paying as much as Rs 110 per kg for air freight from Delhi to Leh. The local administration also brings 250–350 tonnes of vegetables by cargo plane annually for distribution among the civilian population. Meeting the increasing requirement for fresh vegetables for the local populace and the defense personnel deployed in the region is a significant challenge (Angmo et al., 2017).

4.1 Early vegetable introduction

Very few vegetable types were traditionally grown in Ladakh. The most commonly used wild vegetables were *Allium prezewalskianum, Lepidium latifolium, Capparis spinosa, Urtica hyperborean, Lactuca dolicophylla, Fogopyrum tatari* etc (Lamo et al., 2012) before the concept of cultivated vegetable became popular in the region. The adventurous Moorcroft stayed for two years in Ladakh, from September 1820 to September 1822. He stated in his travelogue: "*There is no great variety of vegetable produce in Ladakh, but onions, carrots, turnips, and cabbages are reared in some places during the spring and summer. For winter use the leaves of the cabbages and turnip tops, or sliced turnips, are dried: carraway, mustard, and tobacco are grown in few gardens*" (Moorcroft & Trebeck, 1841). Cunningham (1853) mentioned peas and turnips being grown in Ladakh. The first significant addition to the Ladakhi diet was made by the Moravian missionaries who came to Leh in the last quarter of the nineteenth century. They brought vegetables such as potatoes, spinach, cauliflower, radish, green beans, khol rabi, brussels sprout, and tomatoes (Gabriele et al., 1998).

When the Defence Institute of High Altitude Research (DIHAR), formerly Field Research Laboratory (FRL), was established in Leh in the year 1962, concerted efforts were made in terms of formal R&D to help increase the quantity and diversity of vegetables in the region. Agriculture Research Unit (ARU) was established in the year 1968 to undertake research on vegetables. The vegetable research farm at DRDO-DIHAR continues to be known as ARU even today. The institute developed agro-techniques for growing a variety of vegetables in the region. Seed and seedlings of beans, beetroot, cabbage, cauliflower, carrot, lettuce, spinach, and tomato were distributed among the farmers by DRDO-DIHAR between 1965 and 1969. Quality planting materials of okra, knol-khol, leek, sugar beet, chilies, coriander, cucumber, Chinese cabbage, mint, brinjal, garlic, methi, and pumpkin were distributed in the 1970s. Field trials of capsicum, squash, and Karam Sag began in the late 1970s, and the sale of seedlings began in the early 1980s. The institute distributed 644750, 771825, and 868470 vegetable seedlings to farmers and army units in 1975, 1985, and 1995, respectively (Stobdan et al., 2018). By the year 1980, 45 types of vegetables were being grown in Ladakh. Later the Regional Agricultural Research Station of SKUAST-K, established in 1989, also conducted research and extension to promote vegetable cultivation in the region. Consequently, with the help of the State Agriculture Department and the local non-governmental organizations (NGOs), farmer's training and extension activities were conducted to promote vegetable cultivation in the region. The locally produced vegetables and consumption choices are now far greater than what it was a few decades back.

4.2 Cultural practices and production calendar

Cultural practices and vegetable production calendars in the Ladakh region are different from other parts of the country. Single-cropping is dominant due to long, harsh winters. The cropping season extends from May to September in a major part of the region. The use of vegetable seedlings raised in a passive solar greenhouse is commonly practiced in the region. Vegetable nurseries are raised in greenhouses from 21 March onwards, and seedlings are transplanted in open-field in early May. Nursery raising in the greenhouse extends the crop growing season almost by two months and thus results in obtaining harvesting maturity of the majority of vegetable crops. Hardy crops such as cabbage and onion are transplanted first, and temperature-sensitive crops such as capsicum, brinjal, and cucurbits are transplanted from mid-June onwards. Direct seed sowing of cucumber, pumpkin, and summer squash are also done when black plastic mulch is used (Stobdan et al., 2018). Turnip is among the first crops to harvest in the region from May onwards. Summer squash, knol-khol, and leafy vegetables reached the marketable stage from June onwards. Most crops reach the harvestable stage from late July onwards and are harvested before October. The sub-zero temperature in October is not unusual in the region, and delays in harvesting often result in frost damage (Stobdan et al., 2018).

4.3 Vegetable production data

The vegetable production in Ladakh was approximately 250 tonnes during 1971–72, which increased to nearly 2,500 tonnes in 1981. Ladakh region produced 17,572.2 tonnes of vegetables in 2022 (Leh: 9,071.2 tonnes; Kargil: 8,500.9 tonnes). Most vegetable farms are small, measuring less than 0.2 acres (Stobdan et al., 2018). The area of vegetable production constitutes 4.5 percent of the total agricultural land in the region.

A variety of crops are being grown by the farmers. The crops most widely produced in Leh district are potato (75.6), peas (10.7), onion (3.4), cabbage (3.3), carrot (2.0), and cauliflower (1.8 percent). Preference for crops such as potato, cabbage, onion, and carrot is primarily because of their long-term storage capacity during winter when the region remains cut-off and open-field cultivation is impossible (Stobdan et al., 2018).

4.4 Pest and diseases

Insect pests and diseases are not a major problem in Ladakh. The cabbage aphid (*Brevicoryne brassicae* Linn.) is the most common aphid species that cause serious damage to cabbage, cauliflower, broccoli, and Brussels sprouts. It also attacks Chinese broccoli, Chinese cabbage, radish, and kale. Onion maggot (*Delia antique*) is a serious pest in major onion-growing areas in Ladakh. The pest results in 50-65 percent onion bulb yield loss in severely infested villages (Gupta et al., 2021). Cutworms feed on several vegetables at night, damaging the crop near the ground. The cabbage butterfly (*Pieris brassicae nepalensis*) is a major pest of cabbage, cauliflower, broccoli, and knol-khol.

4.5 Vegetable storage

Due to the freezing winter and remoteness of the region, the people in Ladakh have mastered the art of storing root crops, potatoes, onions, and cabbage. The traditional storage methods are entirely based on the use of local natural resources. The stored vegetables are an important source of nutritious food during the snow-covered period. It helps to break the monotonous consumption of dried vegetables. Despite the modernizing activity in the region, the traditional method of vegetable storage is still playing an important role in the life of the region's people. The three most common vegetable storage methods in the region are *Sadong, Tsothbang,* and *Churches.* An underground storage pit, locally known as *Sadong,* is used for storing radishes, carrots, turnips, and potatoes. The pits vary in size, depending on the quantity of vegetables to be stored. The *tsothbang* is a rectangular structure with a small entrance and a window. The window acts as a ventilator. The structure is made either in the house's basement or as a separate outdoor underground or semi-underground structure. *Churches* method is used for onion storage by hanging them from the ceiling (Ali et al., 2012).

4.6 Characteristic features of vegetable production in Ladakh

Ladakh's topography, agro-climatic conditions, and the cultural ethos of the people living in the region make some of the vegetable production features unique.

(1)   Vegetables are harvested from June to September, out of which the majority of them with bulk production, such as potato, onion, cabbage, carrot, etc., are harvested from late July to September. Therefore, excess vegetable produce is available during this period, while in winter, fresh vegetables are scarce in the region. During winter months, passive solar greenhouse cultivation is the only source of locally produced fresh leafy vegetables. Selected vegetables such as potatoes, radishes, carrots, cabbage, onion, and turnip are stored in underground pits and cellars for their consumption in winter months (Ali et al., 2012).

(2)   Ladakh's climatic condition is ideal for producing cole and root crops, and vegetable seed production.

(3)   A variety of cool- and warm- season vegetables are grown in summer. DRDO-DIHAR has demonstrated

the feasibility of growing 101 types of vegetables in a single season in the same field. The same was recorded in the Limca Book of Records in 2013.

(4)   Growing extra-large sized vegetables is common in Ladakh. DRDO-DIHAR has registered the record of growing extra-large vegetables (pumpkin 35 kg, cabbage 14.2 kg, turnip 4.3 kg, knol khol 3.2 kg, potato 1.1 kg, onion 750 g) in the Limca Book of Records.

(5)   Almost every household in Ladakh owns a naturally ventilated passive solar greenhouse for vegetable production during winter months.

(6)   Vegetable seedlings raised in a greenhouse are commonly used in Ladakh. It extends the cropping season almost by two months.

(7)   Farmers in the region use minimal off-farm inputs. The vegetable farming system focuses on ecological harmony by recycling nutrients and maintaining ecological balances for soil and crop management.

(8)   High crop yield is obtained under a low input system without chemical fertilizer, pesticide, and weedicide. Application of only farm yard manure (25 tonnes per hectare) gives an acceptable yield.

(9)   Standard cultural practices, such as the use of pesticides and weedicides, staking, pruning, raised beds, etc., are not followed. Crops are grown more densely than recommended.

(10)   Disease and insect pest infestation is minimal in the region; hence, the vegetable products are free of chemical pesticides.

(11)   Agriculture production is entirely based on irrigation. Water is the main limiting factor for expanding the area under cultivation.

(12)   Women play a decisive role in producing and selling vegetables in the Ladakhi society.

(13)   Vegetables are generally sold directly to consumers; the army is the main buyer of locally produced vegetables.

DRDO-DIHAR felicitates farmers for growing extra-large organic vegetables

4.7 Competitive advantages of vegetable production in Ladakh

Ladakh region has distinct competitive advantages over the plains due to its topography, agro-climatic conditions, traditional farming system, and remoteness.

Off-season production

Production of fresh vegetables after or before their normal season is referred to as off-season vegetable production. The objective is to produce and supply vegetables to the market during their lean supply period. Ladakh has a distinct competitive advantage in off-season vegetable production due to its agro-climatic condition. Vegetables such as broccoli, cauliflower, cabbage, and peas are being produced in July-September in open field that cannot be produced in the plains during the summer, which makes these vegetables off-season to the plains. However, large-scale vegetable cultivation and marketing of the product as off-season vegetables has not been explored yet.

Organic produce

The market for organic vegetables is currently exhibiting strong growth in India. Ladakh region has several competitive advantages in organic vegetable cultivation. Disease and insect infestation are the least in the region; hence, the vegetable produced is free of chemical pesticides. Due to the inherent traditional farming system and remoteness, the vegetable farming system focuses on ecological harmony through the recycling of nutrients and maintaining ecological balances for soil and crop management. Strong support of the government under various schemes acts as a catalyst in promoting organic farming in the region. This situation presents a huge potential to promote organic vegetable farming for increased and sustainable food production, and enhanced income for the farmers.

Nutrient-dense vegetables

Consumers are becoming more conscious of their well-being, and this has led to a boost in demand for nutrient-dense food items. The high-altitude environment of Ladakh, especially the wide temperature amplitude, high light intensity, and high UV radiation, results in vegetables with higher nutrient contents as compared to those grown in low-altitude regions.

Investigations have been carried out to evaluate the comparative phytochemical profiles of Arugula (*Eruca sativa*) plants grown in Ladakh (3,524 m) vs. Chandigarh (321 m). The study concludes that Arugula grown at high altitudes gets enriched with health-promoting bioactive phyto-compounds compared to plants grown at lower altitudes (Kumar et al., 2022). Similarly, it has been found that the mean content of total phenolics, flavonoids, carotenoids, tocopherols, and ascorbic acid is higher in the microgreens grown under high altitude Leh condition than when grown under low altitude Delhi conditions (Priti et al., 2021). Therefore, there is a scope to promote vegetables being grown in Ladakh as nutrient-dense vegetables.

Seed production

Vegetable seed production is a lucrative enterprise, and the Ladakh region is ideal for seed production. The region has low humidity, bright sunshine, long days, low incidence of disease and insect pest infestation, nearly flat to slightly sloping topography, and well-drained soil, which are favorable for seed production. The high mountains that separate the villages act as natural barriers to cross-pollination. However, vegetable seed production in Ladakh is in the infancy stage.

Growing demand for local vegetable produce

Vegetable production in Ladakh has increased during the last three decades primarily due to technological intervention, access to farm inputs, and growing market demand. Campaigns by various government and non-government organizations have resulted in increased awareness about the importance of growing vegetables. Apart from growing vegetables for self-consumption, supply to the army and tourism industry is a major factor in the region's rapid transformation. Marketing of vegetables, especially potatoes, peas, broccoli, and turnip, to other parts of the country is also being witnessed in recent years.

4.8 The way forward

Significant progress has been made in promoting vegetable cultivation in Ladakh. The major focus has been on growing a variety of vegetables in the summer months, identification of suitable varieties, passive solar greenhouse cultivation in winter months, the early raising of vegetable seedlings, and storage of hardy vegetables during winter months. In the coming years focus should be on the following:

Fresh vegetables in winter: The availability of fresh vegetables during winter months is restricted to leafy vegetables and selected tuber and root crops. The passive solar Ladakh Greenhouse has made it possible to grow cauliflower, broccoli, cabbage, and tomatoes in peak winter months. Further research is required to increase the diversity and quantity of vegetables for winter months. The majority of the greenhouses established in the region are small and designed to meet a family's vegetable requirement. There is a need to tap the potential of the high-tech greenhouses, which can be the new future economy of the region.

Organic certification: Minimal use of off-farm inputs is a characteristic feature of Ladakh agriculture. The majority of the farmers restrain from the use of pesticides and weedicides. Organic certification is required to fetch a higher price for the farm produce. There is a need to link organic farming with the tourism industry.

Organic seed production: The organic food industry is flourishing due to consumers' preference for organically grown produce over conventionally grown crops. As per regulation, organic farmers must use organic seed material if such is available; as a result, there is a frantic search for certified organic seed by growers. Only a few seed companies have organic seed production facilities, so organic seed is in short supply. The agro-climatic condition of Ladakh is ideal for the seed production of selected vegetable crops. The high mountains that separate the villages act as a natural barrier to cross-pollination. However, organic vegetable seed production in Ladakh is in the infancy stage. Focus attention is required towards building human resources for producing high-value organic vegetable seeds.

Advanced production technology: Vegetable growers in Ladakh face many challenges, and there is a need to invest in developing new and innovative production technologies. Improved production technology is needed to enable farmers to remain on the farmland. Plasticulture is a proven technology that needs to be promoted among growers. Farmer training and extension activities are required to achieve increases in productive efficiency. Farm mechanization is required to combat labor scarcity, especially during critical farm activities.

Post-harvest management: Most of the vegetable crops attain the harvesting stage in August and September. Given the perishable nature of the produce, excess produce gets spoils. Growers, therefore, restrict the growing of perishable vegetables in small areas. Storage inadequacies result in limited vegetable production. There is a need to promote food processing industries to make available local produce during winter months. A lack of insufficient technical expertise limits food processing in the region.

Short duration crop: The higher elevation areas, especially above 4000 m asl, have cooler temperatures and a shorter growing season that is often measured in weeks rather than months. Besides, the weather is unpredictable during the growing season, and wild animals often damage crops. There is a need to identify short-duration crops that can be grown under multiple microclimates. Growing vegetables in such places are even more important since most people are semi-nomads, and their diet mainly consists of meat and milk-based food.

Traditional underutilized local crop: The underutilized wild plants that have been traditionally used as vegetables remain neglected. Such plants are nutrition-dense, climate-resilient, economically viable, and able to grow on marginal land. Using such plants takes into account indigenous people's traditional knowledge and cultural identity. These underutilized plants are key to agricultural diversification in Ladakh. Lakhs of tourists who visit Ladakh annually can be the potential targets for popularizing such underutilized plants, thus creating extra income and investment for local people. Research, breeding, and development efforts are needed to tap its commercial potential.

Crop selection for enhanced yield and quality: Most well-designed studies showed that maximizing crop yield and quality, in terms of nutrient density, cannot be achieved. Modern-day agriculture has excelled in pushing crops' physiological limits through intensifying input use. Plant breeders have consistently found ways to overcome constraints to higher yields. However, yield enhancement has come at the expense of nutrient density, organoleptic quality, the environment, plant health, and food safety. A growing body of research shows that organic farming increases the concentration of nutrients in food by 25 percent compared to conventionally grown crops (Benbrook, 2009). However, an effort must be made to balance enhanced crop yield without compromising the nutrient quality. It would be interesting to study whether crop yield and quality can be maximized under the organic production system in the Ladakh region with its unique agro-climatic conditions, especially the long day length, high UV radiation, and cool nights. Potato, the most widely grown vegetable, is a heavy feeder. There is a need for a gradual shift towards high-value crops that requires less manure.

Selecting hybrid and open-pollinated cultivars for yield benefit: Locally adapted, traditional varieties perform equally or even better than the high-yielding and hybrid varieties under environmental stress conditions. Therefore, efforts are being made globally to breed new varieties for sustainable cultivation under sub-optimal growing conditions and a low-input organic system. However, the performance of organic vegetable varieties, compared to conventional high-yielding and hybrid varieties, under different environmental conditions has yet to be intensively researched. Preliminary studies that we conducted at DRDO-DIHAR showed that hybrids perform significantly better in terms of vegetable yield in Leh conditions. Therefore, rejecting the use of high-yielding hybrid varieties under organic systems in Ladakh agro-climatic conditions based on studies elsewhere would be unwise.

Increasing agricultural water productivity: Water is one of the major limiting natural resources for large-scale vegetable cultivation in the region. The future direction and success of Ladakh vegetable cultivation will depend on water availability, particularly the ability to increase water productivity. There is so much to learn from Israel's irrigation practices. The geography and climate of both Ladakh and Israel are not naturally conducive to agriculture. Both regions are largely desert, and the lack of water resources does not favor farming. Israel's agricultural sector is characterized by an intensive system of production stemming from the need to overcome the scarcity of natural resources, particularly water and arable land.

Insect pest and disease management: The incidence of insect pests and diseases in the cold arid region is low. However, the incidence of cutworms, onion maggots, aphids, and cabbage butterflies emerged as the main pest of vegetables in the region, inflicting a huge economic loss on the growers. However, a detailed study on the identification and reporting of the major insect pests and diseases have not been carried out in the region. Correctly identifying diseases and insect pests makes controlling them easier and more effective. Most importantly, it helps identify possible threats to agriculture and the environment. Since the emergence and population dynamics of pathogens and insects are linked to abiotic and biotic factors, it is expected that the crop disease and insect-pest problems of the high mountain region of Ladakh would be different from that of the plains. Therefore, there is a need for well-designed studies for the identification and management of diseases and insect pests in the region.

Breeding for multiple abiotic stresses: Due to the region's topography and agro-climatic condition, the crops growing in the region are subject to multiple abiotic stresses, especially drought, cold and high light intensity. Given the emerging genomic-assisted breeding, it would be fascinating if region-specific crop varieties could be developed with multiple abiotic stress tolerance.

Harnessing niches: Ladakh region has some comparative advantages over the plains. However, the niches have yet to be harnessed to date. Infrastructural support needs to be developed to harness the niches. A variety of vegetables can be marketed as off-season vegetables to the plains. Contract farming needs to be introduced, especially for exotic vegetables on a large scale, to meet the demand of the niche market. There is a need to enhance the capacity of farmers to increase the production and profitability of these vegetables, especially using organic inputs. Off-season vegetables have a lot of potential to bring in additional income as they sell at much higher prices.

Cultivation on barren land: There is great scope to expand the area under vegetable cultivation from the existing 1,163 hectares. With technological advancement, lifting water from the river to irrigate the vast unused barren land is now feasible.

Capacity building: Conscious effort is required to assist vegetable farmers through capacity building in adopting new and innovative production technologies. Farmers need to develop marketing skills to embrace agricultural marketing on their terms in a manner that cuts out intermediaries and connects food producers directly with customers.

Disruptive technologies: The climatic condition of Ladakh is not conducive to large-scale vegetable production. However, advances in artificial intelligence, robotics, and sensing technologies are expected to disrupt today's agriculture. The region needs to look for disruptive technologies that can be developed in-house or adapted elsewhere.

REFERENCES

Ali Z, Yadav A, Stobdan T, Singh SB (2012). Traditional methods for storage of vegetables in cold arid region of Ladakh, India. *Indian J Tradit Knowl.*, 11: 351–353

Angmo S, Angmo P, Dolkar D, Norbu T, Paljor E, Kumar B, Stobdan T (2017). All year round vegetable cultivation in trenches in cold arid trans-Himalayan Ladakh. *Def Life Sci J.*, 2: 54–58

Benbrook C (2009). The impacts of yield on nutritional quality: Lessons from organic farming. *HortScience* 44(1): 12–14

Cunningham A (1854). *Ladakh: Physical, statistical, and historical with notices of the surrounding countries.* Edition 1997. Gulshan Publisher, Gow Kadal, Srinagar, India

Dame J, Nüsser M (2011). Food security in high mountain regions: agricultural production and the impact of food subsidies in Ladakh, Northern India. *Food Sec.*, 3: 179–194

Dar RA, Rather GM (2014). Assessment of magnitude of malnutrition and related health problems in cold desert Ladakh-India. *Eur Acad Res.*, 2(4): 4895–4919

Gabriele R, Angmo P, Dolma T (1998). *Ladakhi kitchen: traditional and modern recipes from Ladakh.* Melong Publications, Ladakh, India

Gupta V, Raghuvanshi MS, Namgyal D, Landol S, Dorjey S, Stanzin J, Tundup P (2021). Bio-efficacy of different insecticides/ bio-insecticides against onion maggot (*Delia antique*) under in-vitro conditions in cold arid region of India. *Pharma Innov J.*, 10(5): 1254–1258

Kumar N, Kaur B, Shukla S, Patel M, Thakur M, Kumar R, Chaurasia OP, Khatri M, Saxena S (2022). Comparative analysis of phytochemical composition and anti-oxidant and anti-inflammatory benefits of *Eruca sativa* grown at high altitude than at lower altitude. *Chem Pap.*, 76: 7759–7782

Lamo K, Akbar PI, Mir MS (2012). Underexplored and underutilized traditional vegetables of the cold arid Ladakh region of India. *Vegetos*, 25: 271–273

Mishra GP, Singh N, Kumar H, Singh SB (2010). Protected cultivation for food and nutritional security at Ladakh. *Def Sci J.*, 60: 219–225

Moorcroft W, Trebeck G (1841). Travels in the Himalayan Provinces of Ladakh and Kashmir in

Peshawar, Kabul, Kunduz and Bokhara from 1819 to 1825. Edited by HH Wilson. Gulshan Books, Srinagar, India

Pelliciardi V (2013). From self-sufficiency to dependence on imported food-grain in Leh district (Ladakh, Indian trans-Himalaya). *Eur J Sustain Dev.*, 2: 109–122

Priti, Mishra GP, Dikshit HK, T. V, Tontang MT, Stobdan T, Sangwan S, Aski M, Singh A, Kumar RR, Tripathi K, Kumar S, Nair RM and Praveen S (2021). Diversity in phytochemical composition, antioxidant capacities, and nutrient contents among mungbean and lentil microgreens when grown at plain-altitude region (Delhi) and high-altitude region (Leh-Ladakh), India. *Front Plant Sci.*, 12: 710812

Stobdan T, Angmo S, Angchok D, Paljor E, Dawa T, Tsetan T, Chaurasia OP (2018). Vegetable production scenario in trans-Himalayan Leh Ladakh region, India. *Def Sci J.*, 3: 85–92

Chapter-5

Popular Fruits of Ladakh: Status and Way Forward

Fruits are an essential part of a healthy diet, and variety is as important as quantity. The rising trend of cancers and cancer-related deaths is a major concern in the Ladakh region. The crude cancer rate in the region is 31.5 cases per lakh per year. Low intake of fruit is significantly associated with cancer in the region. Cancer is highest (58.8 percent) in patients taking less than one fruit serving per week, 24.6 percent in those taking one to four fruit servings per week, and only 16.7 percent in those taking more than four fruit servings per week (Hussain et al., 2019). Therefore, adequate consumption of fruit is required to prevent the development of cancer in the region.

Limited varieties of fresh fruits are available for a short period for consumption in the region. Apricots and apples are the two main fruit crops. Fresh apricots are available during July-September, while locally grown apples are readily available during August-October. Minor quantities of grapes, peaches, cherries, and plums are found in the region. Therefore, most fruit varieties need to be obtained from outside the region through goods trucks from Manali or Srinagar, covering a distance of over 420 km to Leh. During winter, a limited quantity of fresh fruit is brought in by air, paying as much as Rs 110 per kg for air freight from Delhi to Leh.

Ladakh has many natural advantages for temperate fruit production. The region experiences long day hours with high light intensity and relatively warm days with cool nights and low relative humidity from May to October. However, fruit trees are being grown in the region as individual trees or small groups of trees. Standard cultural practices such as pruning, training, manuring, and spacing are not being followed by the majority of the growers. Despite having a huge potential for quality fruit production, the fruit industry in the region is at a nascent stage. We surveyed in 2021 and found that most of the growers do not follow the standard growing practices. Only 11.4 percent of the respondents are growing apple trees in the orchard system. Knowledge about insect pests and diseases is poor in the region. When asked to select the biggest constraints in apple production, 32.6 percent of the respondents selected lack of marketing opportunity, followed by high insect-pest infestation (24.8 percent), insufficient water for irrigation purposes (21.1 percent), non-availability of nursery plants (12.4 percent) and lack of fencing around the field (9 percent) (Dolker et al., 2022).

5.1 Apricot

The apricot (*Prunus armeniaca* L.) is a traditional fruit crop of the Ladakh region, and it is locally known as *Chuli*. A few trees believed to be 100 to 125 years old can be seen in Ladakh today. Except for trial purposes, the region has yet to witness the introduction of cultivars from outside the region. The gene pool is maintained due to geographical isolation and the natural high mountain barrier. The fruit crop is deeply associated with the tradition and culture of the region. Historically, dried apricot was one of the four natural products of Ladakh that were traded with neighboring countries (Cunningham, 1854). *Phating,* the premium quality dried apricots of Ladakh, is popularly known as *Nyari Khambu* in Tibet (Stobdan et al., 2021). *Raktsey Karpo*, apricots with white seed stone, is unique to Ladakh (Angmo et al., 2017). It is the most preferred cultivar for fresh consumption (Naryal et al., 2019a). The oldest written document on *Raktsey Karpo* is found in the travelogue of the adventurous Moorcroft, who stayed for two years in Ladakh from September 1820 to September 1822. He stated in his travelogue: "*This is a small fruit not much larger than a walnut, somewhat flattened at top and bottom, of a glossy skin, and pale yellow colour, inclining to white, which changes to a reddish brown where it faces the sun. The pulp is of the usual consistence next the skin, but becomes softer as it recedes, and next the stone is little thicker than honey in the comb. The whole fruit partakes of the lusciousness of honey, combined with a slight and agreeable bitter, and the flavour is unsurpassed by any variety of apricot I have ever met with. The stone is of a light yellow colour, approaching to white. The trees grow in the Pargana of Ladakh proper, and especially at Saspula. Not far from Lé, on the bank of the river*" (Moorcroft and Trebeck, 1837).

Apricots of Ladakh are known for their quality. However, today, Ladakhi apricots' popularity remains restricted to the region primarily due to a lack of awareness and quarantine restrictions. However, in recent years the apricots of Ladakh have been receiving attention from the government and the industry. The maiden export of Ladakh apricot began in 2021 to Dubai. During the 2022 season, 35 tonnes of fresh apricots were sent to the domestic market outside Ladakh and also exported to Singapore, Mauritius, and Vietnam.

Apricot is classified in Ladakh into two broad categories based on kernel taste. Fruits with a bitter kernel are called *Khante*, meaning bitter, while those with a sweet kernel are called *Ngarmo*, meaning sweet (Targais et al., 2011). The *Ngarmo* is further divided into two sub-groups based on seed stone color. Fruit with white seed stones is called *Raktsey Karpo*, while those with brown are called *Nyarmo*. Apricots with white seed stones are unique to Ladakh and are associated with the sweet kernel, brightly colored fruit with high TSS (Angmo et al., 2017). Oil extracted from the sweet kernel is used for edible purposes either in pure form or mixed with walnut oil. A spoonful of oil is mixed with finely ground roasted barley flour, salted tea, and sugar to prepare a local dish called *Phemar* which is served to guests and during a festive occasion. The bitter kernel is used to extract oil, which has religious, cosmetic, and medicinal values. Even today, the oil is used as hair oil and relieves backache and joint aches. The oil is popular as body and massage oil and is known for penetrating the skin without leaving an oily feel. Applying warm apricot oil mixed with a pinch of common salt on the chest relieves patients of acidity (Targais et al., 2011).

Raktsey Karpo, Ladakh's first GI-tag product

5.1.1 Area and production

The total area under apricot cultivation is 2,612 ha (Kargil: 1,670 ha; Leh: 942 ha) in 2022. Ladakh produces approximately 15,864 tonnes of fresh apricot, making it the largest producer of apricot in the country. The region produced approximately 2,000 tonnes of dried apricot, making it the country's largest producer of dried apricot. Not all parts of Ladakh produce premium-quality dried apricots. Micro-climate plays a vital role in the production of apricots that possess characteristics of quality dried fruits. Generally, villages in a narrow valley below 9,500 feet elevation produce the best quality dried apricots. Villages known for producing premium quality dried apricots (*Phating*) in the Kargil district are Batalik, Chulichan, Darchik, Dargo, Garkhon, Gongma, Gurgurdo, Hardaz, Hordas, Hundurmal, Kargil, Karkichu, Kharul, Manjee, Majidas, Sanjak, Sanatsey, and Shilikchay.

Similarly, in Leh district, the villages known for producing premium quality dried apricots are Achinathang, Biama, Bokdang, Dha, Domkhar-Dho, Hanu Thang, Lehdo, Skurbuchan, Takmachik, Thang and Turtuk (Stobdan et al., 2021).

5.1.2 Competitive advantages of growing apricot in Ladakh

Apricot is one of the few temperate fruit trees unaffected by overproduction, and often premium prices are reported for fresh and processed fruits (Hormaza et al., 2007). Therefore, there is immense scope for Ladakh to emerge as a major apricot-producing region. Apricots of Ladakh are known for their quality. The uniqueness of Ladakh's apricot is evident from the fact that historically the dried apricot was traded with neighboring countries (Cunningham, 1854).

***Raktsey Karpo* is unique to Ladakh:** *Raktsey Karpo*, apricots with white seed stones, are unique to Ladakh. It has not been reported anywhere else in the world. White seed stone phenotype is associated with a sweet kernel, brightly colored fruit, and exceptional sweetness (Angmo et al., 2017). It is the most preferred cultivar by consumers for fresh consumption in terms of sweetness, juiciness, aroma, flesh color, and overall appreciation (Naryal et al., 2019a). The white seed coat phenotype can be explored as a distinguishing feature of high-quality apricots of Ladakh (Angmo et al., 2017). The cultivar got the Geographical Indication (GI) tag in December 2022 and became the first product of Ladakh to get the GI tag.

Late fruit ripening: Apricots of Ladakh are harvested between mid-July and early September. In comparison, apricots in Spain attain maturity in mid-May and late June (Ruiz and Egea, 2008), and those in Greece and America are harvested from May to June (Drogoudi et al., 2008). Apricots in Anatolia, Turkey, are harvested in late June and early July (Asma and Ozturk, 2005). Those from Lake Van Region, Turkey, are harvested in late July to early August (Balta et al., 2002). In India, the fruit is harvested from Kashmir, Kullu, and Uttarakhand in May-June. Therefore, apricots of the Ladakh region have a distinct competitive advantage as it does not coincide with the main apricot season in the market.

Extended marketing season: Apricots are available for fresh consumption for a short period in the market due to their perishable nature. Therefore, there is a great demand for fresh apricot for an extended period. The harvesting season of apricots in Ladakh extends from mid-July to early September, primarily due to the altitudinal variation. Genetic diversity also contributes to extending fruit harvesting. Within an orchard at DRDO-DIHAR Leh, we recorded a 39-day difference in harvest between the earliest and the late genotypes (Naryal et al., 2020). Harvesting spanning more than a month is a crucial element given the short shelf life of fresh apricot fruit. Therefore, apricots of Ladakh can be marketed over an extended period (Stobdan et al., 2021).

Brightly colored fruit: Visual appeal is an important trait demanded by the fresh fruit market. Consumers are attracted to orange skin and intense blush. We conducted a survey and found that consumers appreciate fruits with medium-orange skin and prefer fruit with a large blush area. Apricots of Ladakh have 313.8±101.0 mm² blush area and light to dark orange flesh (Naryal et al. 2019a). Therefore, brightly colored apricots of Ladakh have a competitive advantage in the domestic and international markets.

Sweetness: A lack of sugar or sweetness in purchased apricot fruit is among the most common consumer complaints (Moreau-Rio and Roty, 1998). Apricots of Ladakh are sweeter. The sugar content in terms of total soluble solids (TSS) ranged from 11.0–37.9ºBrix with a mean value of 23.9±5.7ºBrix. Fruit sweetness has a direct relationship with altitude of the growing area. For every 100 m increase in elevation, the fruit TSS increased by 1.2ºBrix. *Raktsey Karpo* has significantly higher TSS (28.1±3.8ºBrix) than cultivars with brown seed stone (Angmo et al., 2017). In comparison, the TSS of apricots in Turkey ranged from 11–27ºBrix (Asma and Ozturk, 2005), and 12.71–20.0ºBrix in Pakistan (Ali et al., 2011), 10.6–16.3ºBrix in Spain (Ruiz and Egea, 2008), 8.7–22.4ºBrix in France (Gurrieri et al., 2001), and 12.3–15.8ºBrix in Italian apricots (Leccese et al., 2012).

Fruit rich in sorbitol: Sorbitol is a sugar alcohol that is used as a sweetening agent in various food products, such as sugar-free sweets and chewing gum. It has 60 percent of the sweetness of sucrose, with one-third fewer calories. It does not contribute towards dental caries and is suggested to be helpful to people with diabetes. Apricots of Ladakh are rich in sorbitol. Fruit sorbitol contents increase with increasing altitude. *Raktsey Karpo* cultivar contains significantly higher sorbitol, followed by fruits with a brown coat with the sweet kernel and fruits with a brown coat with the bitter kernel (Naryal et al., 2019c).

Small fruit: Apricots of Ladakh are small. Fruit weight ranged from 7.1–53.7 g with a mean weight of 21.6±9.3 g. In comparison, the fruit weight of 21 apricot cultivars collected from Canada, the Czech Republic, Ukraine, and the USA ranged between 28.1–77.7 g with a mean weight of 42.44 g (Vachůn, 2003). We found an inverse relationship between altitude and fruit weight of apricots of Ladakh. For every 100 m increase in elevation, the fruit weight decreases by 0.5 g (Naryal et al., 2000). Therefore, the small fruit size of the apricot of Ladakh is due to genetic factors and the region's high elevation. Therefore, fruit weighing 25–30 g has a comparative advantage since it is easy to consume such fruits.

Organic: Chemical fertilizers, pesticides, and chemical sprays are not used throughout the growing cycle. The fruits are more nutritious and tastier. Besides, it ensures a healthier and safer environment.

5.1.3 New initiatives

Considering the huge potential and holistic growth of apricots in Ladakh, several new initiatives have been taken in recent years. Under the government's One District One Product (ODOP) scheme, the apricot has been chosen for the Kargil district. The scheme provides the framework for value chain development and alignment of support infrastructure. The Ladakh Tourism Department is organizing the 'Apricot Blossom

Festival' from 2021 onwards. It is a celebration of the beautiful apricot blossoms during spring. It helps in creating awareness and promoting apricots of Ladakh. The maiden export of Ladakh apricot began in 2021, and fresh apricots are sent to the domestic market outside Ladakh. Defence Research Development Organisation (DRDO) and the Administration of Union Territory of Ladakh established a model apricot processing plant in 2022 to demonstrate complete supply chain management of apricot. It is the first project in Ladakh operated and maintained under the Government Owned Contractor Operated (GOCO) model. *Raktsey Karpo*, apricots with white seed coat, got the Geographical Indication (GI) tag in December 2022 and became Ladakh's first GI-tagged product.

5.2 Apple

Apple (*Malus domestica* Borkh.) is a traditional fruit crop of the Ladakh region, locally known as *Kushu*. The region represents a great wealth of indigenous apple germplasm that varies in color, size, flavor, and texture. The mention of apples in Ladakh is found in the travelogue of the adventurous Moorcroft, who stayed in Ladakh for two years, from September 1820 to September 1822. He stated in his travelogue: "*The apple-trees are also numerous, and of several varieties: some of them are engrafted, but the greater number are wildings: they bear freely, endure great cold and intense heat, require little rain, and are very rarely attacked by disease. The ordinary fruit is of the middle size, rather oval than round, of very regular shape, and of great beauty and variety of colour: it is very juicy, and of an agreeable, though not very decided flavour, and the pulp is light, without being at all woolly. They are ripe in September, and are kept in very good preservation through the winter*" (Moorcroft and Trebeck, 1837).

The total area under apple cultivation is 946 ha (Leh: 572 ha; Kargil: 374 ha) in 2022. Ladakh produces approximately 5,191 tonnes of fresh apples, making it the region's second most important fruit crop, after apricot. Several native cultivars are grown in the region, and *Thra*, *Mongol*, and *Karkechu* are the three most popular cultivars. Other lesser-known cultivars are *Marpo*, *Yayon*, *Bong*, and *Skyurmo* (Dolker et al., 2021). The local cultivars are ripe in August but do not store well. However, the fruit of Delicious exotic cultivars ripe in late October and stored well for 4–5 months by the traditional storage method. Because of late ripening and prolonged fruit storage quality, there is an increasing demand for Delicious cultivars in the region.

Ladakh has many natural advantages for apple production. The region experiences long day hours with high light intensity and relatively warm days with cool nights and low relative humidity from May to October, thus making apple cultivation favorable in the region. However, apples are being grown in the region as individual trees or small groups without using chemical fertilizers and pesticides. A survey conducted in Leh district in 2021 revealed that most farmers in the region are small apple growers, and 76.6 percent of respondents have less than 20 apple trees. Despite having a vast potential for quality apple production, the region's apple industry is in the nascent stage (Dolker et al., 2022).

Reports suggested climate change has affected apple production in major apple-growing regions in India. Rising winter temperature affects apple production in lower altitude areas of Kullu, Sirmaur, Shimla, Mandi, and Solan in Himachal Pradesh (Gautam et al., 2014). Besides the yield loss, the quality of fruit produced in these areas has also degraded, and orchardists have to resort to chemical sprays for color development (Sharma et al., 2017). The consequences of climate change have resulted in the shifting of apple cultivation from lower to higher altitudes (Gautam et al., 2014). Because of the changing scenario, the high-altitude Ladakh region has the potential to emerge as an ideal place growing quality organic apples (Dolker et al., 2023).

5.2.1 The uniqueness of Ladakhi apples

Karkechu **cultivar**: *Karkechu* is a special apple cultivar of the Ladakh region known for its fruity aroma. In some parts of Ladakh, it is also known as *Saspol* or *Saspolo*. On maturity, the aroma of the fruit can be smelled from a distance. If kept in a closed room, the entire room smells fantastic. The volatile compound profile of *Karkechu* is significantly different from other apple cultivars. Given its unique characteristics, an application has been filed for GI tagging of *Karkechu*.

Late flowering: The exotic Delicious apple bloom in mid to late May in Leh condition (elevation 3,331 m). Late flowering is essential in protecting from damage caused by spring frost and, therefore, a desirable trait in regions experiencing spring frost. The native *Thra* cultivar blooms 12–24 days earlier than the exotic cultivars. Early flowering in the native cultivar may be an adaptive mechanism for early completion of the fruit development stages before the onset of cold weather in the region (Dolker et al., 2021).

Late fruit ripening: Exotic apple cultivars in Leh condition (elevation 3331 m) are harvested in late October. In comparison, apples in mid hills of Himachal are harvested between mid-July to late-September (Sharma et al., 2017), while those from the mid-hills Uttarakhand are harvested between late July to late-August (Kishor et al., 2018). Late ripening of the exotic cultivars has a competitive advantage. The harvesting season does not coincide with other major apple-producing regions (Dolker et al., 2023). Besides, late ripening is favorable for overwintering fruit storage for winter consumption in the region.

Intense red skin fruit: The visual appearance determines the perceived quality of red or bicolored apples. In general, red fruits are the most preferred. Even with adequate size, poor fruit color is generally associated with poor visual consumer acceptance. Several expensive methods are being used to improve the fruit color. However, under the climatic conditions of Ladakh, the red apple cultivars naturally formed intense red skin fruits due to the prevailing low temperature. Ladakh region also experienced intense sunlight and high ultraviolet radiation that further contributed to the development of red skin color in apples. Therefore, the intense red-colored apples of Ladakh have a competitive advantage in the domestic and international markets (Dolker et al., 2023).

Sweetness: Sweet taste is an important driver of consumer preference, and TSS is the most convenient predictor to determine the sweet taste in apples. The fruit TSS of Delicious cultivars grown in Leh range from 11.1–15.9ºBrix with a mean value of 13.2±1.8ºBrix (Dolker et al., 2023). In comparison, the value ranged from 11.4–14.5ºBrix in apples grown in Kullu, Himachal (Kumar et al., 2019), 10.4–15.9ºBrix in mid-hills of Himachal (Sharma et al., 2017), 11.3–14.7ºBrix in mid-hills of Uttarakhand (Kishor et al., 2018), 12.2–14.1ºBrix in Kashmir (Mushtaq et al., 2018). Higher TSS in apples grown in Ladakh may be due to high altitude environmental conditions and organic production system.

Organic: Most apple growers do not use chemical fertilizers, pesticides, and chemical sprays. The market for organic apples is currently exhibiting strong growth, and Ladakh has a competitive edge in organic apple production.

5.3 Seabuckthorn

Seabuckthorn (*Hippophae rhamnoides* L.) is an ecologically and economically important thorny shrub widely distributed in Ladakh. The female plant bears red, orange, or yellow berries on its two-year-old thorny branches. Seabuckthorn berry is one of the most nutritious fruits with medicinal properties. Ladakh remains the primary site for natural Seabuckthorn resources, with over 70 percent of the country's total area (13,000 ha) under Seabuckthorn. Once considered a thorny menace, Seabuckthorn is now seen as a means for the region's sustainable development (Stobdan et al.,

2017). I had the opportunity to lead a study entitled 'Value chain analysis of Seabuckthorn in Leh Ladakh' commissioned by the Ministry of Agriculture and Farmers Welfare, GoI. Based on the report, Seabuckthorn is now included as a horticultural activity under the Mission for Integrated Development of Horticulture (MIDH) scheme in Ladakh, Himachal, Uttarakhand, Arunachal, and Sikkim since Apr 2018. Farmers and processors in Ladakh are getting incentives from the Horticulture Department for growing and processing Seabuckthorn. Centre of Excellence on Seabuckthorn is sanctioned under the MIDH scheme, and the building is near completion at Nimoo, Leh Ladakh.

5.4 Peach

The peach (*Prunus persica* (L.) Batsch) is a traditional minor fruit crop of the Ladakh region; locally, the fruit is known as *Tra-kushu*. The total production is just 34 tonnes in 2022. The fruit of native cultivars are small in size and lack attractive skin color, and thus peach has not emerged as a popular fruit crop of the region. However, Ladakh has many natural advantages for peach production. We conducted systematic studies on the introduction and performance evaluation of exotic peach cultivars. The exotic cultivars bloom from late April to early May, and the fruit is ripe from late August to early September. The cultivar Redhaven ripened earlier, followed by Glohaven and Suncrest. The average fruit weight of the Suncrest, Redhaven, and Glohaven are 87.6, 65.1, and 117.8 g, respectively. An 8-year-old Suncrest peach tree yields 56.4 kg. The fruit yield is significantly low in cultivars Glohaven (14.4 kg per tree) and

Redhaven (5.6 per tree). The average fruit sweetness, in terms of TSS, of Suncrest, Redhaven, and Glohaven are 10.1, 13.2, and 9.0°Brix.

Growing peaches in Ladakh has a comparative advantage over other fruit crops. It has a relatively short juvenile period of 2–3 years compared to most other fruit tree species that require 5–10 years. We recommend the cultivar Suncrest for the region because of its high yield, large fruit size, and attractive red fruit skin color. Peach can be marketed as off-season fruit to other parts of the country if grown on a large scale. It is ripe from late August to early September in Ladakh. Peaches are usually harvested from April to July in most peach-producing regions, such as Punjab, Haryana, and the adjacent areas of Western Uttar Pradesh, Uttarakhand, and Himachal Pradesh. The intense red-colored peaches of Ladakh will have a competitive advantage in the domestic and international markets.

5.5 Mulberry

Mulberry (*Morus alba* L.) is a traditional minor fruit crop of the Ladakh region, locally known as *Oosey*. It is found at 2700-3300 m asl on hilly slopes along the rivers. In most mulberry-growing countries, particularly in India and China, mulberry is used for its foliage to feed the silkworm. However, in Ladakh, mulberry is grown as a fruit crop (Bajpai et al. 2014a), and fresh and dried fruits in small quantities are sold in the local market. Fruiting occurs in July, and the fruit color varies from white, purple, or red (Bajpai et al. 2014a, 2014b). Fruit weight ranged from 0.6 to 3.6 gm. The fruit sweetness in terms of TSS varied from 12.6 to 38.1ºBrix.

5.6 Watermelon

Watermelon (*Citrullus lanatus*) is traditionally not grown in the Ladakh region, and local requirement is transported from outside the region. Locally watermelon is known as *Zagon.* The increasing popularity and market share of watermelon in the region have led us to conduct studies on the introduction and performance evaluation at DRDO-DIHAR, Leh. Field trials began in 2014 at an elevation of 3,344 m. Successful experimental trials on varietal selection and black plastic mulch enable the farmers in the region to grow watermelon in open field conditions. The best-performing cultivar (Bejo-2000) under black mulch yielded an average of 57.4 tonnes per hectare of marketable fruit. Watermelon has been grown commercially from 2016 onwards. Direct seed sowing has resulted in growing watermelon in far-flung villages in the region. The variety Bejo-2000 is recommended and widely cultivated in the region due to its high yield (Angmo et al., 2019).

Watermelons are usually grown in open fields from February onwards and harvested from April to June in India. However, in Ladakh, it matures in August and September and thus can be marketed as off-season fruit to other parts of the country. Growing watermelons is highly profitable. Due to off-season and organic produce, it fetches a higher price. The fruit sells at Rs 35-80 per kg in the region, and farmers earn a net income of Rs 12,37,000 per hectare (Lakdan and Stanzen, 2017).

5.7 Muskmelon

Muskmelon (*Cucumis melo* L.) is traditionally not grown in the Ladakh region. There is no report of studies related to growing muskmelon in high-altitude arid areas, particularly in the region beyond 3,000 m asl. However, due to the increasing popularity of muskmelon in the region, we at DRDO-DIHAR conducted studies on growing organic muskmelon in the open field.

The marketable yield of five varieties ranged from 5.0 ± 0.3 to 18.8 ± 1.7 tonnes per hectare using black plastic mulch. The best-performing cultivar (Pusa Madhuras) under black mulch yielded an average of 17.3 tonnes per hectare marketable fruit under open-field conditions. The success of the research trial is attracting farmers to grow these fruit types. However, fruit cracking is a major problem in the commercial cultivation of muskmelon. Pusa Madhuras, Khushbu, Patasha, and Rustam are suitable for the Ladakh region. Muskmelons are usually grown in open fields from mid-January to mid-February and harvested from April to June in major muskmelon-growing areas in India (Singh et al., 2022). However, in Ladakh conditions, it is harvested in August and September and thus can be marketed as off-season fruit to other parts of the country (Angmo et al., 2018).

5.8 Sun melon

Sun melon (*Cucumis melo* var. *inodorous*) is highly relished because of its attractive fruit with a unique flavor and sweet taste. We conducted studies with the Indian Agricultural Research Institute (IARI) and found that sun melon can be successfully grown as a cash crop under organic farming in open-field conditions in Ladakh. Varietal selection and black plastic mulch are key for growing sun melons. The best-performing cultivar (Pusa Sarda Melon Hybrid-3) under plastic mulch yielded an average of 25-35 tonnes per hectare marketable fruit. The success of the research trial is attracting farmers to grow sun melon in the region. We recommend direct seed sowing in the open field in mid-May to ease the widespread reach in far-flung areas. Pusa Sarda, Pusa Sunehari, and Pusa Sarda Melon Hybrid-3 are recommended. Sun melons are usually grown in the net-house condition in the north Indian plains. However, in Ladakh condition, it ripens in August and September in open field and thus can be marketed as off-season fruit to other parts of the country. The fruit has a long shelf life and can be stored at room temperature for 15–20 days. The fruit can be transported over a long distance without any damage. It has an attractive color; thus, the fruit has potential for export markets.

5.9 Strawberry

Most strawberry varieties are short-day, which cannot be grown in Ladakh conditions. However, long-day and day-neutral varieties are successfully being grown in the region. The ideal time for planting runners is mid-March to mid-April. It comes into flowering in May to June in open-field conditions, while in the greenhouse, it starts flowering by the end of March to April. Fruits are ready for harvesting 5–6 weeks after flowering, and 3–4 pickings are taken at weekly intervals. A well-managed crop in an open field gives an average yield of 60 quintals per hectare. It is a perennial plant that survives the freezing temperature during winter if covered with mulch. However, the crop is not popular among the farmers but has potential as a cash crop in the region. Gorella, Addie, Hera, and Confetura are long-day varieties that perform well in Ladakh conditions. Recommended day-neutral varieties are Brighton, Tieoga, Selva, and Fern (Dwivedi et al., 2001).

5.10 The way forward

Fruit crops have received little attention as compared to vegetables in the region. A handful of researchers and extension workers are involved in research and extension services related to fruit crops. The primary focus has been on promoting Delicious apple cultivars and apricot drying. The following priority areas need attention in the coming years:

Product differentiation and branding

Promoting local products as special and unique is one of the most effective methods of bringing extra value to Ladakh horticultural produce. Branding also helps with consumer recognition of products, ensures higher prices for the same product, and directly accelerates value addition. The unique fruit varieties, such as the GI-tagged *Raktsey Karpo* variety of apricot and the *Karkechu* apple of Ladakh, possess unique properties. Establishing trusted brands and changing consumer perception of Ladakh's unique fruit varieties is vital to bringing these products to market. Organic certification is essential for product differentiation and branding. There is a need to promote Ladakh's special and unique products in national and international agricultural exhibitions, which will undoubtedly increase the visibility and promotion of the brand and sales of the produce. The future of the fruit industry in Ladakh depends on branding, and the product should not be marketed as a commodity.

Area expansion

If Ladakh has to emerge as an important player in the fruit industry at the national or international level, the current area under fruit crops must increase manifold. Large tracts of vacant arable land can be brought into cultivation with minimal environmental impact. There is immense scope to expand the area from the existing 3,671 hectares under fruit crops. Fruit crop plantation has a relatively long gestation period ranging from 6–7 years from the planting date, and it involves substantial capital investment. Public-private partnership (PPP) would be essential for area expansion under fruit crops. The focus should be on planting premium and improved cultivars, such as *Halman* and *Raktsey Karpo*, in the case of apricot.

Increasing agricultural water productivity

Horticultural crop production in Ladakh is entirely based on irrigation, and water is one of the major limiting natural resources. Hot and dry conditions result in high evaporation rates. The soil is primarily sandy loam with significantly less moisture-holding capacity. Hence, the crops need more water with higher frequency. Water scarcity in the region could be addressed through measures that improve water productivity and increase land productivity. Various technologies are available for enhanced operation, better management, and efficient irrigation water use. The future direction and success of the Ladakh fruit industry will depend on water availability. Ladakh needs to learn from Israel's irrigation practices.

Orchard system

Fruit trees are planted around the farmers' houses or field boundaries. Well-managed orchards are hard to find in the region. Planting fruit trees under the orchard system augments yield per hectare and simplifies farming operations. Therefore, there is a need to advocate growing fruit trees under a well-managed orchard system. Establishing an orchard is a long-term investment and deserves critical planning.

Fruit nursery

The supply of quality nursery plants of known cultivars is vital for promoting fruit crops in the region. Nursery plants of exotic cultivars of apple, pear, peach, cherry, and other fruit crops can be procured from outside the region. However, in the case of apricot and Seabuckthorn, the availability of quality nursery plants is a major concern. The nursery plants need to be raised locally. Commercial nursery raising needs immediate attention.

Insect pest and disease management

The incidence of insect pests and diseases is low. The high mountain natural geographical barrier restricts the spread of insect pests and diseases in the region. However, the incidence of codling moth (*Cydia pomonella* L.) on apples and aphids on apricot is the region's major problem, inflicting a substantial economic loss on the growers. Besides, infestation by woolly apple aphid, *Eriosoma lanigerum* (Hausmann), is seen in a few places. Gummosis is a significant problem for stone fruits in the region. Focus attention is required for the integrated management of major insect pests and diseases in the region without affecting the organic status of the products. There is a need to create awareness about managing insect pests and diseases in the region.

Establishment of marketing infrastructure

Marketing infrastructure for fresh fruit is at the nascent stage. Most apricots and apples are locally consumed and lack an established marketing system for trading outside the region. There are opportunities for fresh as well as processed products from Ladakh to be sold in down-country supermarkets. There is a need to establish a marketing infrastructure for fresh and processed fruits from Ladakh.

Value-added products

A small quantity of processed products, such as jam, juice, and seed oil, is produced by a few privately-owned firms. Despite the 15,864 tonnes of apricots produced annually, less than 50 tonnes of apricot pulp are being used for product development and are locally consumed. Approximately 40–60 percent of the apricot produced in the region is wasted (Stobdan et al., 2020). In the case of

Seabuckthorn, less than 5 percent of the berry is harvested primarily because of a need for an integrated value chain and market-oriented sustainable system. Over 90 percent of Seabuckthorn harvested in Ladakh is sold after primary processing without any value addition (Stobdan and Phunchok, 2017). Therefore, a huge opportunity lies in the processing and value-addition of agricultural and horticultural produce.

Capacity building

Conscious effort is required to assist growers through capacity building in adopting standard cultural practices and innovative production technologies. Most growers do not perform well-established cultural practices like pruning and training trees. There is a need for an intense awareness campaign on the need for training, pruning, and other cultural practices. Farmers must be taken on exposure tours to show orchards where trees are regularly trained and pruned. Front-line demonstrations need to be conducted in every village. Besides, there is a need to develop marketing skills among the growers to embrace agricultural marketing on their terms in a manner that cuts out intermediaries and connects food producers directly with customers.

Conservation of genetic resources

Conservation of indigenous genetic resources is important to national policies and international treaties. Native germplasm can be a novel donor parent for agronomically important traits such as cold and drought tolerance. Ladakh is rich in genetic resources of apricot, apple, and Seabuckthorn. Trees yielding low-quality fruits are often uprooted to make space for cultivars to produce high-quality fruits. Besides house constructions, road widening works often result in uprooting fruit trees. There is a need to conserve the rich genetic resources of the region.

Diversification to high-value crops

Apricots and apples are the two main fruit crops of Ladakh. Preliminary studies have shown that the climatic condition of Ladakh offers scope to grow a variety of other high-value temperate fruit trees. We are conducting experimental trials on growing exotic cherries, peaches, and prunes at DRDO-DIHAR. More high-value fruit varieties need to be evaluated to increase the fruit basket. Encouraging farmers to diversify to high-value crops offers great scope for generating significantly higher employment and helping in increasing farmers' income.

Breeding for multiple abiotic stresses

Due to the region's topography and agro-climatic condition, the crops growing in the region are subject to multiple abiotic stresses, especially drought, cold, and high light intensity. Given the emerging genomic-assisted breeding, it would be fascinating if region-specific fruit varieties can be developed with multiple abiotic stress tolerance.

Whole genome sequencing of elite cultivars

The elite fruit cultivars, such as the *Raktsey Karpo* apricot and *Karkechu* apples, exhibit unique characteristics. Whole genome sequencing of the elite cultivar is important to understand genome evolution and important traits. Further, it will aid in identifying alleles that could be readily combined to drive rapid varietal development. Plant variety identification is vital in agricultural systems, and genome sequencing is the accurate variety identification method.

Disruptive technologies

The climatic condition of Ladakh is not conducive to the large-scale growing of fruit trees. The region must look for disruptive technologies such as artificial intelligence, robotics, and sensing technologies.

REFERENCES

Ali S, Masud T, Abbasi KS (2011). Physico-chemical characteristics of apricot (*Prunus armeniaca* L.) grown in Northern areas of Pakistan. *Sci. Hortic.*, 130: 386–392

Angmo P, Angmo S, Upadhyay SS, Targais K, Kumar B, Stobdan T (2017). Apricots (*Prunus armeniaca* L.) of trans-Himalayan Ladakh: Potential candidate for fruit quality breeding programs. *Sci. Hortic.*, 218: 187–192

Angmo S, Dolkar D, Dolkar P, Kumar B, Stobdan T (2019). Growing watermelon in high altitude trans-Himalayan Ladakh. *Natl Acad Sci Lett.*, 42: 379–382

Angmo S, Stobdan T, Chaurasia OP, Katiyar AK (2018). Growing muskmelon (*Cucumis melo* L.) under a low-input system in arid trans-Himalayan Ladakh, India. *Def Life Sci J,* 3: 96–99

Asma BM, Ozturk K (2005). Analysis of morphological, pomological, and yield characteristics of some apricot germplasm in Turkey. *Genet Resour Crop Evol.*, 52: 305–313

Bajpai PK, Warghat AR, Dhar P, Kant A, Srivastava RB, Stobdan T (2014a). Variability and relationship of fruit color and sampling location with antioxidant capacities and bioactive content in *Morus alba* L. fruit from trans-Himalaya, India. *LWT-Food Sci Technol.*, 59(2): 981–988

Bajpai PK, Warghat AR, Sharma RK, Yadav A, Thakur AK, Srivastava RB, Stobdan T (2014b). Structure and genetic diversity of natural populations of *Morus alba* L. in trans-Himalayan Ladakh region. *Biochem Genet.*, 52: 137–152

Balta F, Kaya T, Yarilgaç T, Kazankaya A, Balta MF, Koyuncu MA (2002). Promising apricot genetic resources from the Lake Van Region. *Genet Resour Crop Evol.*, 49: 409–413

Cunningham A (1854). Ladakh: Physical, statistical, and historical with notices of the surrounding countries. Gulshan Publisher, Gow Kadal, Srinagar, India

Dolker T, Katiyar AK, Chaurasia OP, Stobdan T (2022). Farming practices, knowledge, and constraints in apple production in Ladakh: a survey. *J Food Agric Res.*, 2: 59–69

Dolker T, Dolma T, Kumar Deepak, Sharma NC, Chaurasia OP, Stobdan T (2023). Growing organic exotic apples in trans-Himalayan Ladakh in climate change scenarios. *Proc Natl Acad Sci India Sect B Biol Sci.*, DOI: 10.1007/s40011-023-01489-w

Dolker T, Kumar D, Chandel JS, Angmo S, Chaurasia OP, Stobdan T (2021). Phenological and pomological characteristics of native apple (*Malus domestica* Borkh.) cultivars of trans-Himalayan Ladakh, India. *Def Life Sci J.*, 6: 63–68. doi: 10.14429/dlsj.6.15726

Drogoudi PD, Vemmos S, Pentelidis G, Petri E, Tzoutzoukou C, Karayiannis I (2008). Physical characters and antioxidant, sugar, and mineral nutrient contents in fruit from 29 apricots (*Prunus armeniaca* L.) cultivars and hybrids. *J Agr Food Chem.*, 56: 10754–10760

Dwivedi S, Attrey DP, Paljor E (2001). Strawberry cultivation in Ladakh. Field Research Laboratory

Gautam HR, Sharma IM, Kumar RA (2014). Climate change is affecting apple cultivation in Himachal Pradesh. *Current Sci.*, 106: 498–499

Gurrieri F, Audergon J-M, Albagnac G, Reich M (2001). Soluble sugars and carboxylic acids in ripe apricot fruit as parameters for distinguishing different cultivars. *Euphytica*, 117: 183–189

Hussain S, Ali M, Jeelani R, Abass M (2019). Cancer burden in high altitude Kargil Ladakh: Ten-year single centre descriptive study. *Int J Cancer Treat.*, 2(2): 4–10

Kishor A, Narayan R, Brijwal M, Attri BL, Kumar A, Debmath S (2018). Performance of some new apple cultivars for yield and physico chemical characters under mid-hill conditions of Uttarakhand. *Indian J Hort.*, 75: 8–14

Lakdan S, Stanzen L (2017). Economic analysis of watermelon based on production system in Trans-Himalaya region of Ladakh. *J Pharmacogn Phytochem.*, 6(6): 2602–2604

Leccese A, Bartolini S, Viti R (2012). From genotype to apricot fruit quality: the antioxidant properties contribution. *Plant Food Hum. Nutr.*, 67: 317–325

Mushtaq R, Pandit AH, Ali MT, Javaid K, Sharma MK, Singh A, Bashir D (2018). Evaluation of two exotic apple varieties on M9T337 for growth and quality attributes under Kashmir conditions. *Int J Curr Microbiol App Sci.*, 7: 2828–2832

Naryal A, Acharya S, Bhardwaj AK, Kant A, Chaurasia OP, Stobdan T (2019c). Altitudinal effect on sugar contents and sugar profiles in dried apricot (*Prunus armeniaca* L.) fruit. *J Food Compos Anal.*, 76, 27–32

Naryal A, Angmo S, Angmo P, Kant A, Chaurasia OP, Stobdan T (2019a). Sensory attributes and consumer appreciation of fresh apricots with white seed coats. *Hortic Environ Biotechnol.*, 60: 603–610

Naryal A, Dolkar D, Bhardwaj AK, Kant A, Chaurasia OP, Stobdan T (2020). Effect of altitude on the phenology and fruit quality attributes of apricot

(*Prunus armeniaca* L.) fruits. *Def Life Sci J.,* 5: 18–24

Ruiz D, Egea J (2008). Phenotypic diversity and relationships of fruit quality traits in apricot (*Prunus armeniaca* L.) germplasm. *Euphytica*, 163: 143–158

Sharma DP, Sharma HR, Sharma N (2017). Evaluation of apple cultivars under sub-temperate mid-hill conditions of Himachal Pradesh. *Indian J Hort.,* 74: 162–167

Stobdan T, Dolkar P, Chaurasia OP, Kumar B (2017). Seabuckthorn (*Hippophae rhamnoides* L.) in trans-Himalayan Ladakh, India. *Def Life Sci J.,* 2: 46–53

Stobdan T, Namgial D, Chaurasia OP, Wani M, Phunchok T, Zaffar M (2021). Apricot (*Prunus armeniaca* L.) in trans- Himalayan Ladakh, India: Current status and future directions. *J Food Agric Res.,* 1: 86–105

Stobdan T, Phunchok T (2017). Value chain analysis of Seabuckthorn (*Hippophae rhamnoides* L.) in Leh Ladakh, Ministry of Agriculture and Farmers Welfare, Govt of India

Stobdan T, Chaurasia OP, Wani M, Phunchok T, Zaffar M (2020). Improvement of apricot farming in Ladakh, India' for Union Territory of Ladakh. Administration of UT Ladakh, India

Targais K, Stobdan T, Yadav A, Singh SB (2011). Extraction of apricot kernel oil in cold desert Ladakh, India. *Indian J Tradit Knowl.,* 10: 304–306

Vachůn Z (2003). Variability of 21 apricot (*Prunus armeniaca* L.) cultivars and hybrids in selected traits of fruit and stone. *Hortic Sci.*, 30: 90–97

Organic apples of Ladakh: potential fruit with a competitive advantage

Peach cultivar Suncrest: a potential off-season fruit with a competitive advantage

Chapter-6

Blossoms in the Desert: Prospects of Floriculture in Ladakh

Floriculture is emerging as the most diversified and potential component in the horticulture industry, both at the national and international levels. It fetches a higher price as compared to fresh fruits and vegetables. The domestic demand for flowers is increasing exponentially, especially in the metros and larger cities (Anumala and Kumar, 2021). The climatic condition of Ladakh is ideal for the off-season production of cut and loose flowers; and the production of seeds and bulbs. The prevailing low temperature, high photosynthetic irradiance, and UV rays are known to cause intense flower color. Water deficiency, common in Ladakh, also causes flowers to turn darker. Therefore, the flowers in Ladakh develop an intense color, an important determinant that directly influences their commercial value.

6.1 Early history

Not much is known about the early introduction of imported flowers in Ladakh. However, way back in 1978–79, agro-practices for growing a variety of flowers such as antirrhinum, aster, calendula, California poppy, candytuft, carnation, chrysanthemum, cineraria, dahlia, dianthus, gladiolus, hollyhock, marigold, petunia, roses, sweet william, and zinnia were standardized at DRDO-DIHAR. The institute played a significant role in the beautification of homes and army units by distributing a large number of flower seedlings. During 1978–79, about 5 lakh flower seedlings were distributed to the locals. Efforts were made to keep flowering plants alive in winter by keeping the plants in trenches without supplementary heating. For the first time, the flowering of chrysanthemums was recorded in December 1979. *Thuja compacta* and Eucalyptus were kept alive in trenches for two winters.

The maiden effort in developing agro techniques for the commercial cultivation of gladiolus was made in 1994 by DRDO-DIHAR. The State Horticulture Department supplied corms to the farmers at subsidized rates. Later the State Agriculture Department took up the project, and commercial cultivation of gladiolus became popular in the mid-1990s (Dwivedi et al., 2002). However, the project continued for a short time due to poor air connectivity and logistic constraints. A dedicated floriculture farm was established in 2020 by the Agriculture Department, UT Ladakh, at Thiksey village near Leh town to promote floriculture. The farm is becoming a major attraction to the locals.

Lupin in bloom in the high mountain Ladakh

6.2 Cultural practices and production calendar

The cultural practices and production calendars followed in Ladakh differ from other parts of the country. Single-cropping is dominant from May to September. Manuring and soil preparation is done in April. Seedlings are raised in passive solar greenhouses in the last week of March and are transplanted in the open field in early May. Irrigation is done by flooding at three days intervals during initial plant establishment followed by five days at later stages. About 20 to 25 irrigations are required from transplanting to seed harvesting.

6.3 Scope and opportunities in commercial floriculture

Floriculture provides tremendous scope and opportunities as a commercial business venture with high market value. It is a low-volume, high-value business. Scope and opportunities for Ladakh ranged from growing crops such as off-season cut and loose flowers, dry flowers, bulb and seed production, and extraction of essential oil and pigments. The floriculture market is wider than just the domestic market in big cities. The potential of the local market should be considered. Many tourists visit Ladakh, and flowers are in great demand. Besides, the export potential can also be explored with well-established supply chain management.

6.3.1 Cut flowers

A majority of cut flowers in India are grown under controlled climatic conditions. However, Ladakh's climatic condition is ideal for producing off-season cut flowers in open fields. The flowers in Ladakh develop an intense color, an important determinant that directly influences their commercial value. The prevailing low temperature, photosynthetic irradiance, and high UV-B cause intense flower color. Water deficiency, common in Ladakh, also causes flowers to turn darker.

Lilium: Lilium is an important cut flower grown in Ladakh. It is among the top ten cut flowers traded in the world market. Its attractive shape, size, wide range of color, long case life, and exotic appearance makes it very appealing. Commercial cultivation began in Ladakh recently. In August of 2022, the first consignment of lilium, consisting of approximately 800 flower sticks, was shipped from Ladakh to Delhi under CSIR Floriculture Mission for the first time. Growing cut flowers in the open field can increase farm income. However, the market linkage is the deciding factor for the success of the commercial cultivation of lilium. The lilium varieties are generally of three types, namely, Asiatic, Oriental, and Longiflorum hybrids. Based on trials conducted at DRDO-DIHAR, Leh, several lilium varieties have been found suitable for cultivation in the cold-arid conditions of Ladakh (Dwivedi et al., 2002).

Asiatic lilium: Alaska (White), Amsterdam (Orange Red), Elite (Orange), Gran Paradiso (Orange Red), Jessica (Dark Orange), Kinks (Orange), Nove Cento (Yellow), Pasadena (Dark Yellow), Yellow Giant (Yellow).

Oriental hybrid lilium: Alhambra, Amanda, Atlantis, Cascade, Kissporoof, Meditaranne, Marco Polo, Star Fighter, Star Gazer, White Merostar.

Longiflorum hybrids: Avita, Snow Queen, White Fox, White Satin

Gladiolus: Gladiolus is a lovely flowering plant, available in many colors and sizes. Gladioli typically reach between 2 and 4 feet in height. It produces elegant flower spikes with orchid-like blooms. They are spectacular in a vase and are suitable for growing in flower gardens, container gardens, and even vegetable gardens. Growing the flower from corms measuring 8–10 cm in diameter is easy. The corms are planted in late April to early May in Leh condition. It generally takes 100–120 days for flowering, and thus late planting after May is not recommended. The climatic condition of Ladakh results in high-quality spikes with more florets and superior flower colors. The flower stalks reach the cutting stage in mid-July to mid-September and thus have a competitive advantage as off-season cut flowers. In comparison, gladiolus from the plains reached the market from

November to March. The five recommended varieties for Ladakh are American Beauty, Friendship, Princess Margaret Rose, White Prosperity, and Souvenir. Insect-pest infestation is not a major problem in the region (Dwivedi et al., 2002). However, occasional thrips infestation and fungal diseases are seen.

6.3.2 Loose flower

The loose flowers dominate the domestic market and are predominantly grown under open fields. Marigolds dominate the Indian market, and the major producing states are Madhya Pradesh, Karnataka, Gujarat, and Andhra Pradesh (Anumala and Kumar, 2021). The main flowering season of marigolds is mid-October and Feb-March (Malviya et al., 2022). Ladakh region has the competitive advantage for producing off-season loose flowers. Marigold in Ladakh is in bloom from July to September, grown under an organic system. Similar is the case with most other loose flowers. The potential of sending loose flowers to metros and larger cities is yet to be explored.

6.3.3 Flower bulb production

The flower bulb sector produces and trades bulbous and tuberous plants, rhizomes, and root tubers. The climatic condition of Ladakh is favorable for the production of flower bulbs. The low incidence of insect pests and diseases is an added advantage for the region. However, the sector still needs to be addressed, and the region has immense scope for flower bulb production.

6.3.4 Seed production of flowering annuals

Seed production of flowering annuals is a lucrative enterprise, and the Ladakh region is ideal for seed production. The region has low humidity, bright sunshine, long days, low incidence of disease and insect pest infestation, nearly flat to slightly sloping topography, and well-drained soil, which are favorable for seed production. The high mountains that separate the villages act as natural barriers to cross-pollination. Flower seeds have been produced for over five decades for raising seedlings and their subsequent distribution to local and army units at DRDO-DIHAR. However, commercial seed production has yet to be explored to date. In India, flowering annuals are majorly grown as rabi season crops under sub-tropical climatic conditions, which are highly suitable for commercial seed production. It is mainly practiced in the northwestern plains where the crops flower in March-April and seed sets in April-May (Anumala and Kumar 2021). However, in Ladakh, it flowers in June-August and sets seed in August-September. Therefore, the region can be explored for off-season seed production.

6.3.5 Wild rose on commercial scale

Wild rose (*Rosa webbiana* Wall. ex Royle), locally known as '*Sia*', is drought-tolerant and grows even in extremely high-altitude areas in Ladakh. It needs to be planted on a large scale for hips and petals.

REFERENCES

Dwivedi SK, Paljor E, Attrey DP (2002). Lilium in Ladakh. Field Research Laboratory, Leh Ladakh

Dwivedi SK, Paljor E, Attrey DP (2002). Gladiolus in Ladakh. Field Research Laboratory, Leh Ladakh

Malviya AJ, Vala M, Mankad A (2022). Recent floriculture in India. International Association of Biologicals and Computational Digest. 1: 1–8.

Anumala NV, Kumar R (2021). Floriculture sector in India: current status and export potential. *J Hortic Sci Biotechnol.*, 96(5): 673–680

Roses in bloom in the cold desert

Chapter-7

The Organic Ambition of Ladakh

In 2019, the Ladakh Autonomous Hill Development Council (LAHDC), Leh, took its first step to embrace organic farming. After the visit of a team headed by the Hon'ble Chief Executive Councillor, LAHDC Leh, to Sikkim, a Study Group was constituted in January 2019 to examine all aspects of organic farming in the Leh district. The Study Group comprises elected representatives, officers from the Agriculture and Allied departments of LAHDC Leh, and scientists of Leh-based research institutes. DRDO-DIHAR and Agriculture Departments were given the task of leading the Study Group and submitting the report to LAHDC Leh. After several rounds of meetings with all stakeholders, the Study Group submitted its report to LAHDC Leh with the remarks, '*It is feasible to achieve the target of certifying Leh as an organic district by 2025*'. A Special General Council meeting of the LAHDC Leh was held on 9 March 2019, and the Council unanimously declared to become a fully organic district by 2025. The document 'Mission Organic Development Initiative of Ladakh: Policy, Strategy, and Action Plan' prepared by the Study Group was approved for implementation. Later, LAHDC Leh constituted a committee, which I headed, comprising scientists from Leh-based research institutes and officers from Agriculture and allied departments to prepare a project proposal to certify the Leh district as organic by 2025. The project proposal was submitted for funding when Ladakh became a Union Territory the same year. The then Lieutenant governor, Shri R.K. Mathur, launched the Rs 500-crore organic farming project named 'Mission Organic Development Initiative of Ladakh' on 4 July 2020 for both Leh and Kargil districts. A memorandum of understanding (MoU) was signed on 13 July 2021 with Sikkim State Organic Certification Agency (SOCCA) for organic certification.

We planned organic certification of the entire Leh district in three phases. In the first phase, 34 revenue villages were targeted, where chemical fertilizers and pesticides have not been used during the last three years or beyond. In the second phase, it was decided to cover 40 revenue villages where chemical fertilizer and pesticide use was less than five tonnes per year. Organic certification of the remaining revenue villages, which use chemical fertilizer above five tonnes per year, was planned in the third phase.

7.1 Why does Leh Ladakh embrace organic farming?

Subsistence farming is traditionally practiced, and the region was self-sufficient until the 1960s. However, it has now been widely accepted that the current agricultural system cannot help the farmers sustain on small and marginal land. To cope with the changing scenario, agrarian reforms are unavoidable. Without a deep-rooted vision for the future, there is a risk that Ladakh could be overwhelmed by the forces of change and modernization. Therefore, the declaration to certify Ladakh as organic is a timely measure to save the agriculture system in the region. Ladakh remained organic by default based on the inherent traditional farming system and remoteness, providing a solid foundation for developing organic agriculture. Farmers mostly practice mixed farming, where livestock is reared as an integral part of the system for food and manure. The use of chemical fertilizers is among the lowest, and the use of plant protection chemicals is also low. Thirty-four villages in the Leh district do not use inorganic fertilizers. This situation presents a huge potential to promote organic farming for increased and sustainable food production, and for enhanced income for the farmers.

When the declaration was made in 2019 to become fully organic, farmers, extension workers, and researchers witnessed many negative impacts of conventional agriculture. The general public became increasingly aware of the ill effects of chemical fertilizers and pesticides, which threatened the region's fragile ecosystem. Farmers often complain of loss of soil fertility, hardening of soil, low crop yield, and the need for frequent irrigation. The farm products of Ladakh are considered on par with that of the rest of the country, which makes the product a 'commodity' rather than a 'premium produce.' There is a general perception that 'local' produce of Ladakh is significantly healthier than imported ones. Studies have shown that organic farming is more suitable for families with smallholder family farms. In the Leh district, most households have small land holdings, and 49.4 percent have less than 0.5 hectare of land. The total agricultural land of Leh district was just 10,223 ha, i.e., just 0.2% of the geographical area; hence, it is manageable to certify the district as organic. Only 533 tonnes of chemical fertilizers were used in the district during 2017–18. While the national average fertilizer consumption during 2016–17 was 123 kg per ha, it was only 52 kg per ha in the Leh district. The good news was that Leh district had not witnessed an increasing trend in fertilizer use during the last 28 years (1990–2018). The district has a livestock population of 2,80,251, excluding poultry, which was sufficient to meet the organic manure requirement for the entire district.

The incidence of insect pests and diseases is low, and the intensity of pesticide use in the Leh district was very low compared to the national average. While the national average consumption of pesticides during 2014–15 was 0.29 kg per hectare, it was only 0.096 kg per hectare in the Leh district. The district has no privately owned chemical fertilizer and pesticide sale outlets. Chemical fertilizer is being made available to the farmers only through the Cooperative Department of LAHDC, Leh. Similarly, small quantities of pesticides are made available only through State Agriculture and Horticulture Departments. Therefore, it was easy for the administration to regulate chemical fertilizers and pesticides. High mountains geographically isolate each village, and there is minimal risk of contamination and risk of the spread of insect pests and diseases. Being a tourist destination, promoting organic farming is expected to bring quick economic returns. Agro-ecotourism is a growing niche sector globally, and this market mechanism has considerable potential in Ladakh. Therefore, conversion to organic was much easier and more appropriate for the district than later.

7.2 Farmers support organic farming

The key to the success of organic farming in the region lies in farmers' support and strong political will. Soon after the LAHDC, Leh passed a resolution on 9 March 2019 to certify Leh as a fully organic district by 2025, we conducted a survey to gain insights into farmers' attitudes towards adopting the organic policy. Key findings of the survey have been taken into consideration while preparing the project proposal. The survey covers five major agricultural-producing villages at varying distances from Leh town. The findings of the survey are summarized as follows:

Attitude towards the use of chemical fertilizer: The survey suggested that farmers of the region generally accepted that chemical fertilizer affected both soil and human health. Most farmers (60.7 percent) believed chemical fertilizer affects soil health, while 25.4 percent believed it affects human health adversely. A significant percentage of the farmers (10.7 percent) believe that chemical fertilizer affects the quality and taste of the farm produce.

Attitude towards the use of pesticides: The surveyed farmers were very critical of pesticide use. Farmers (96.7 percent) are against the use of pesticides. Most farmers (66.7 percent) believed it kills other living organisms, while 33.3 percent thought it affects human health adversely. The survey suggested that most farmers (76 percent) are against pesticides in Ladakh. In comparison, 12.6 percent are against the LAHDC's abrupt decision to stop the sale of pesticides in the region entirely.

Attitude towards adopting organic farming: Most farmers (83.6 percent) feel that Ladakh should adopt organic farming, while 5.5 percent believe that conventional agriculture should be continued. It emerged that the majority (93 percent) of the respondents would adopt organic farming if they get a 30 percent more price for their organic produce, while only 1.6 percent were against the adoption of organic farming even if they get a premium price for their produce.

Farmers' support for LAHDC decision: The survey suggested that the majority of the farmers (86.3 percent) support the decision of LAHDC Leh to certify Leh as a fully organic district by 2025. Only 6.6 percent are against the decision, while 7.1 percent of the respondents are indecisive.

Farmers' concerns in shifting to fully organic: Farmers also expressed their concerns about the adoption of organic farming. Fodder availability was most farmers' prime concern (41.5 percent). A significant section of the farmers (25.7 percent) was concerned that their crop productivity would reduce with conversion to organic farming. A considerable percentage (21.3 percent) of the surveyed farmers believed that the availability of substitutes for chemical fertilizer is their prime concern. For a small percentage (6 percent) of the respondents, insufficient livestock was their prime concern in adopting organic farming. A small section (1.6 percent) of the respondents believed that managing insect pests under the organic farming system was their prime concern.

7.3 Progress made so far

The year 2023 marks five years since the declaration was made, and Ladakh still has two years to meet the deadline to become 100 percent organic. There has been a slow start in the launch of the organic project. Mission Organic Development Initiative of Ladakh was launched on 4 July 2020 amid the global COVID-19 pandemic. However, despite the slow start and the pandemic, significant progress has been made due to farmers and the government's strong support and strong political will. As of June 2023, an area of 5232 hectares is organically certified under Phase-1, and the villages in Phase-2 will likely be certified by 2025. The organic certification of villages under Phase-3 may have to wait two more years beyond the original deadline. Current data shows a reduction of 48.6 percent in the use of chemical fertilizer in the Leh district compared to 2017. The district has not used any pesticides or herbicides from 2020 onwards. The Ladakh-based research institutes have also reoriented their research works as per the need of the region. After the declaration, we at DRDO-DIHAR stopped using chemical fertilizers and pesticides at our research farms. In studies my students and I have conducted, we found high crop productivity, above the national average, under the organic farming system. Working with LEHO in Takmachik village, we could demonstrate that farmers can get excellent watermelon yield under an organic system.

7.4 Going fully organic : the way forward

With the current pace of development, Ladakh will be a fully organic Union Territory by 2027. Accelerating conversion to organic farming requires increasing investment in region-specific research, education, and innovation. Considering the limited resources available and the size of the rural communities, Ladakh must prioritize growing only those crops commercially for which the region has a competitive edge. The GI tag for *Raktsey Karpo* variety of apricot and the maiden export of apricots in 2021 is an eye-opener for the people of Ladakh. The region is known for its high-quality apricots, apples, and off-season vegetables that can generate significant income. The scope also exists for high-quality seed production. It is too tempting for the farmers to take the easy way out and resort to conventional agriculture. The most effective single thing that can be done to encourage farmers to continue with organic farming is to support them in marketing their organic products at a premium price. Marketing has been a major concern of the farmers. Ladakh's domestic market is a small and challenging place to do business. There is a need to look beyond the domestic market. In recent years, the fresh apricots of Ladakh fetch Rs 100 per kilogram directly to the farmers. The buyers expressed willingness to pay Rs 150 per kilogram if the fruit is organic certified. The availability of organic matter in sufficient quantities is a major concern in supporting organic farming. Mixed farming and the revival of the traditional *Rarzee* and *Barzee* systems are crucial to ensuring the local availability of organic inputs.

Takmachik: An organic village in Ladakh

REFERENCES

Ladakh Autonomous Hill Development Council (LAHDC), Leh Ladakh (2019). Project entitled 'Mission for Organic Development Initiative of Ladakh'

Ladakh Autonomous Hill Development Council (LAHDC), Leh Ladakh (2019). Ladakh organic policy document 'Mission for Organic Development Initiative of Ladakh: Policy, Strategy, and Action Plan'

Passive Solar Greenhouse for Growing Vegetables Under Extreme Conditions of Ladakh

Protected cultivation is a unique and specialized form of agriculture wherein the microenvironment surrounding the plant body is controlled fully or partially as per plant need during their period of growth. The intent is to grow crops where otherwise they could not survive by modifying the natural environment to prolong the harvest period, shorten the time to harvest, increase yields, improve quality, enhance the stability of production, and make products available when there is no outdoor production (Wittwer and Castilla, 1995).

Crop production in the trans-Himalayan Ladakh region is limited by three major environmental constraints: short-growing length in terms of available heat units (HUs), inadequate precipitation, and poor soil fertility (Dolma et al., 2023a). The most determinant factor is the climate. The annual mean maximum and minimum temperature during 2018–2022 is 13.4ºC and 0ºC, respectively. Leh town's hottest month is July (25.8ºC), and January is the coldest (-13.0 ºC). Only four to five months of the year are pleasant for growing crops. Mid-May marks the beginning of the cropping season, while mid-September marks the end of the harvest. Early sowing or transplanting often results in high mortality of seedlings due to spring frost, and late harvesting after mid-September often results in freezing injury. Therefore, limited quantities of vegetables, especially cole and root crops, can be produced in the open field during summer. Vegetables are harvested from June to September, out of which the majority of them with bulk production, such as potato, onion, cabbage, and carrot, are harvested from late August to September. Therefore, excess vegetable produce is available during this period, while in winter, fresh vegetables are scarce in the region. Low temperature is one of the most important abiotic factors that restrict the optimal production of warm-season vegetables in summer. In open field, warm-season crops such as brinjal, capsicum, okra, and bitter gourd give very low yields; thus, farmers do not grow these crops on a large scale.

Meeting the increasing demand for fresh vegetables at an affordable price in this remote mountain area is a formidable challenge. During summer, the vegetables are shipped by truck across the Himalayas, with passes as high as 5300 m asl, covering the distance of Manali to Leh (430 km) or Srinagar to Leh (420 km). However, during winter, a limited quantity of fresh vegetables is brought in by air, paying as much as Rs 110 per kg for the air freight from Delhi to Leh. Fresh vegetables are 2.7-fold costlier in Leh compared to Delhi (Angmo et al., 2019). Bringing fresh vegetables by air is not a viable option as these vegetables are beyond the reach of the common people because of the exorbitant price. Besides, such vegetables freeze before they reach the consumers in far-flung villages.

8.1 Early history

The first greenhouse in Ladakh was established in 1964 at the Defence Institute of High Altitude Research (DIHAR), formerly Field Research Laboratory, as an attempt to produce vegetables during winter (Stobdan, 2015). The greenhouses established during the 1960s to 1970s are made of glass fixed on wooden frames. A few of these greenhouses can still be found in use, for instance, at DRDO-DIHAR experimental station at Ranbirpura, Thiksey. However, the greenhouse with glass covering did not reach the common people. The glass was expensive, and it was a daunting task to transport the fragile material from outside the region. Besides, it requires a costly and sturdy structural system.

The first plastic-covered greenhouse was erected in 1955 in England (Orzolek, 2017), and the use of plastics for greenhouse applications reached Ladakh in the late 1960s. An innovative low-cost greenhouse based on trench warfare was conceived in the late 1960s at DRDO-DIHAR, and vegetables were grown in trenches covered with polyethylene sheets (Angmo et al., 2017). Kerosene heaters were used as supplementary heating at night in peak winter.

The first successful attempt to grow leafy vegetables in winter without supplemental heating was made in 1978–1979. Seeds of spinach, lettuce, spring onion, Karam Sag, and mint sown in September reach the harvestable stage from December till March. The roses in the trench bloomed in November–December. The first attempt to grow warm-season vegetables in winter was made the same year at DRDO-DIHAR. Seedlings of pumpkin, cucumber, kakri, and brinjal were transplanted in July in the trench. The trench was covered with translucent plastic during the day from September onwards, and a black polyethylene sheet was placed on top of the translucent sheet at night to retain heat inside the trench. The plants did not survive from November onwards due to severe cold. Growing mushrooms in winter was attempted during 1979–1980 in a specially designed solar house. The attempt was successful, but a kerosene heater was used at night to heat the solar house.

The widespread use of passive solar greenhouses began in the 1980s when the Ladakh Ecological Development Group (LEDeG), a Ladakh-based volunteer organization, pioneered the construction of Ladakhi polyhouse, passive solar greenhouses with mud walls on three sides (north, east, and west). The greenhouses have a glass cover on the south-facing side, but later the glass was replaced with a polyethylene sheet. The greenhouse with mud walls and polyethylene cover became trendy due to its ease of construction and higher heat retention capability at night. Several modifications have been made to the greenhouse design by various agencies and are named Polyench, GERES greenhouse, LEHO greenhouse, LREDA greenhouse, and SKUAST-II. However, the mud wall on three sides, the use of polyethylene sheet as a cladding material, and angled wooden roof on the north wall are the common features in all these greenhouses (Angmo et al., 2019). The low-cost greenhouse and trench greenhouse was taken up in a big way by the Agriculture Department and other volunteer organizations.

A team of officials from Ladakh visited China in 2007, and the first prototype of a Chinese passive solar greenhouse was established at Gophuk Farm of the Agriculture Department. The greenhouse was found to be superior to the traditional Ladakhi polyhouse for growing leafy vegetables in winter. DRDO-DIHAR later developed Ladakh Greenhouse, an improvised passive solar greenhouse wherein cauliflower, cabbage, broccoli, tomato, and mushroom can be grown in winter months, otherwise impossible in the traditional passive solar greenhouses. The Ladakh Greenhouse is being established on the farmers' field on a large scale by the Agriculture Department, UT Ladakh, from 2020 onwards.

8.2 Objectives of greenhouse cultivation in Ladakh

The overall objective of greenhouse cultivation in the Ladakh region is to modify the natural environment to achieve optimal production of crops. The primary emphasis is on growing vegetables in winter, extending the effective harvest period, early crop harvest, enhancing crop yields, diversification towards high-value crops, alleviating hidden hunger, increasing farm income, reducing CO_2 emission, women empowerment, and providing new opportunities to agri-entrepreneurs.

Growing vegetables in winter: Self-sufficiency in food is an important issue for the region. Filling the gap between the demand and supply of fresh vegetables from November to June is difficult. During winter, the only means for bringing fresh vegetables is by air, which costs as much as Rs 110 per kg for air freight. Therefore, growing vegetables in a greenhouse is the only viable option to meet the increasing demand for fresh vegetables in winter at a reasonable price.

Extending the effective harvest period: The harvesting period of short-duration crops commenced from May onwards. Turnip is among the first crops to harvest. Crops such as knol-khol, summer squash, and leafy vegetables reach the marketable stage from late June onwards. However, most crops reach the harvestable stage from July to September. Due to the sub-zero temperature in October, crops are harvested by mid-September (Stobdan et al., 2018). Growing crops under solar greenhouse can extend the harvest period from early spring to late winter. Greenhouse cucumber and okra are harvested in May, and cauliflower and cabbage in February (Angmo et al., 2020a, 2020b).

Early harvest: Early crop harvest provides better marketing opportunities to the farmers and fetches higher prices for their produce. Greenhouse tomatoes reached the harvestable stage 19 days earlier than open-field (Angmo et al., 2021). Similarly, capsicum in the naturally ventilated greenhouse is harvested 17 to 29 days earlier than in the open field (Angmo et al., 2022). Cucumbers in a solar greenhouse require just 37 days for the first harvest compared to 85 days in the open field (Dolma et al., 2023a).

Enhancing yields: Low temperature is one of the most important abiotic factors restricting the optimal production of warm-season vegetables. Ladakh's climatic condition is unfavorable for field production of warm-season crops such as capsicum, brinjal, and cucurbits. Marketable yield is low in open-field conditions, and most local market requirements are transported from outside the region. However, growing warm-season vegetables in a greenhouse significantly increase the marketable yield. Cucumbers in solar greenhouse yield 3.05 ± 0.24 kg against 1.03 ± 0.16 kg fruit per plant in open-field (Dolma et al., 2023a). The marketable yield of capsicum inside a solar greenhouse is 4.5 to 4.8-fold higher than that of the open field (Angmo et al., 2022). Similarly, a 1.8-fold higher tomato marketable yield is obtained inside a naturally ventilated greenhouse than in an open field (Angmo et al., 2021).

Diversification towards high-value crops: Greenhouse provides opportunities for high-value crops not grown in the open field due to unfavorable conditions. Cherry tomatoes and colored capsicum, for instance, grow well under a naturally ventilated greenhouse. Similarly, growing flower and ornamental crops in place of vegetables can potentially maximize net profit for the growers.

Alleviate hidden hunger: The availability of locally grown fresh vegetables is restricted to summer months; therefore, there are seasonal differences in the dietary intake of vegetables. Seasonal shortfall and low dietary diversity among the local populace led to micronutrient deficiencies, a phenomenon that has been described as 'hidden hunger' (Dame and Nüsser, 2011). A study reported that 35% of the people in Ladakh suffer from malnutrition-related diseases. The most prevalent diseases are anemia (8.7), followed by night blindness (7.3), scurvy (6.7), beriberi (6), pellagra (3.7), and rickets (3 percent of the surveyed subjects. (Dar and

Rather, 2014). Therefore, the all-year-round growing of vegetables in the greenhouse is key to alleviating hidden hunger in the region.

Increasing farm income: Farmers can grow three cycles of crops in a year inside a solar greenhouse against a single cycle of the crop in the open field. This ensures the farmers produce a large quantity of fresh vegetables and round-the-year income. It is estimated that a farmer earns approximately Rs one lakh from a single Ladakh Greenhouse (18.3 m length, 7.3 m width), which is more than the overall farm income of most farmers from their agricultural land.

Reduce CO_2 emission: Ladakh produces a limited quantity of vegetables. Most of the local market vegetable requirement is transported from outside the region. A survey conducted in 2015 suggested that approximately 21.4 ml of diesel is burned to transport one kilogram of fresh vegetables by truck from a nearby town at a 450 km distance (Angmo et al., 2018). This leads to the emission of vehicular pollution in the fragile environment. During winter, a limited quantity of fresh vegetables is brought in by air, contributing to carbon emissions. Growing fresh vegetables in a greenhouse reduces the transportation of vegetables from other parts of the country through road and air cargo, thereby contributing towards reducing carbon emissions and achieving the vision of Carbon Neutral Ladakh.

Women empowerment: In Ladakh, women are the backbone of the agriculture and allied sectors. Women have a major role in the production and marketing of greenhouse vegetables. Production of greenhouse vegetables led to their income generation and empowerment.

Agri-entrepreneurs: Greenhouses provide new opportunities for agri-entrepreneurs for fresh organic vegetable production at a commercial scale to meet the huge vegetable demand. Many unemployed educated youths are looking for region-specific entrepreneurship opportunities.

Residential proximity to greenhouse vegetables: The location of vegetable production closer to consumers is a major benefit. It ensures higher availability and accessibility to fresh vegetables. Besides, it minimizes or eliminates vegetable spoilage due to long-distance transportation and freezing injury.

8.3 Factors affecting the performance of a passive solar greenhouse

Several factors affect the performance of a passive solar greenhouse. The important factors include the covering/ glazing material, composition of the greenhouse wall, size of the greenhouse, and the use of a thermal blanket. Ever since the large-scale adoption of passive solar greenhouses began in the 1980s, the three-side greenhouse wall has been given prime focus to improve greenhouse performance. The greenhouses have been designed keeping in view their utility in winter and thus remain unused in summer.

8.3.1 Covering material

Choosing the right greenhouse glazing or covering material is the most crucial decision in selecting the design for a greenhouse. The covering material used influences the productivity and performance of a greenhouse. The selection of glazing material is critical in attaining an optimal environment, particularly relating to the temperature, solar radiation intensity, and the type of solar energy that reaches the plants inside the greenhouse.

When selecting a glazing material or greenhouse covering, many factors need to be considered. The life of the material, its strength, weight, initial cost, light transmittance, thermal conductance, maintenance issues, and flammability are all critical. There is a range of different glazing options for greenhouses, and they all have advantages and disadvantages when compared. Choosing the right glazing for greenhouses in Ladakh depends primarily on the environment and budget.

Plastic films have been the most prevalent greenhouse covering in Ladakh. Over 99 percent of the existing greenhouses are covered with plastic films. However, in recent years we found that polycarbonate is superior to plastic films. Polycarbonate has good thermal efficiency. It is tough and can bear heavy snow loads. It produces a slightly diffused light which helps prevent burning/scorching the plant. The material lasts up to 15–20 years before replacement is required. However, polycarbonate sheets are prone to scratching, inhibiting solar transmission. The twin and tri-wall flutes can attract moisture, molds, and bugs if not sealed sufficiently in the frame. Installation of polycarbonate sheets requires professionals and is more expensive than polythene sheets.

8.3.2 Greenhouse wall

Tunnel-type greenhouses, covered with polythene sheets, are unsuitable for the Ladakh region. The night temperature often drops to less than -10°C during peak winters. Therefore, a greenhouse with walls on the east, west, and north sides, where the amount of incident solar energy is limited, is strongly recommended for the region. Apart from the load-bearing, the wall's primary function is its ability to absorb solar energy during the day and release the conserved heat inside the greenhouse during nighttime. Single or double-wall mud brick or rammed earth wall is common. Insulation materials such as straw, sawdust, and dry leaves are stuffed between the cavities of the double wall of the greenhouse.

There are various materials to choose from for constructing a greenhouse wall, each with advantages and disadvantages. The suitable material for greenhouse walls in conditions like Ladakh depends primarily on its thermal properties, cost, local availability, and ease of construction. The use of sun-dried mud brick and stone are the two viable options for the construction of greenhouse walls. Sun-dried mud brick wall has been the material of choice for constructing greenhouses walls mainly due to its availability on the site. Sun-dried mud bricks are available in proper shape and size and are easier to work with than stone. However, using agricultural soil to produce mud bricks depletes and degrades fertile agricultural fields. Besides degrading the agricultural soil, sun-dried mud walls have several other limitations. Rain and high humidity inside the greenhouse damage a mud brick wall and thus needs frequent repair and maintenance. Sun-dried mud brick walls are too weak to support additional fixtures such as retrofitted shelves, trays, and hooks to hang baskets inside the greenhouse.

Stones are in abundance in Ladakh. However, its use for greenhouse wall construction is limited. Stone walls have several advantages over traditional sun-dried mud brick walls. The stone's inherent properties are durability, strength, and resistance against the environment's vagaries. Rainwater and high humidity inside the greenhouse have no impact on the stones. There is no shrinkage and swelling from extreme atmospheric variations. Stone walls require little maintenance, repair, or replacement compared to other building material options. Most importantly, stone walls can absorb more heat during day time and release the same at night compared to mud walls. However, the labor-intensive mining of boulders and dressing them into the desired shape necessary for constructing a wall is a time-consuming process that slows the construction speed.

Composite walls comprise a number of layers of different materials with different properties and thicknesses. A composite wall comprising a 1-foot thick outer wall made out of sun-dried mud brick and a 2-feet thick inner wall constructed out of stone masonry stuffed with a glass wool insulation material in-between is a viable option for Ladakh. We found that the extra mud brick wall and the insulation material increase the night temperature by approximately 1.5°C inside the greenhouse during peak winter compared to a 2-feet thick single stone wall. A double wall system with an inside wall constructed out of stone masonry provides all the advantages of a stone wall. Besides, it provides better thermal insulation as compared to a single wall. However, the construction cost of a double wall is higher than a single wall. More time and materials for construction are required, and thicker walls occupy more land.

8.3.3 Greenhouse size

The length, span, and height of a passive solar greenhouse are essential factors that determine the temperature and humidity distribution inside the greenhouse and also influence the cost of the structure. Large greenhouses can achieve a more uniform, stable, and ultimately superior growing environment for the crop. The large greenhouse has greater air volume, which helps slow the abrupt temperature changes. The larger air volume takes longer to heat up or cool down. However, large greenhouses are expensive to build. Due to the small land holding, it is difficult for an ordinary farmer to construct a large greenhouse. Besides, large greenhouses are difficult to maintain, leading to underutilization.

The majority of the farmers in Ladakh have a preference for low-cost small-size passive solar greenhouses. We conducted a door-to-door survey of the Leh district and found that most (56.3 percent) farmers have a small greenhouse (9.8 m x 5.5 m). Only 3 percent of the farmers have medium size (19.8 m x 7.3 m), while a small section (1.8 percent) has a large size (30.5 m x 7.0 m) passive solar greenhouses (Angmo et al., 2019). Preferences for small greenhouses are due to small land holdings and socio-economic conditions (Dolma et al., 2023b). The general perception among people that smaller greenhouses are warmer than large greenhouses is not true. In a recently concluded study, we found that a large solar greenhouse has a more positive effect on the growth and yield of cabbage and cauliflower during winter in Ladakh. The study used two similar passive solar greenhouses differing in length and span. The length and span of the large greenhouse were 27.4 m and 8.2 m, while that of the small greenhouse was 9.8 m and 5.5 m, respectively. The study found that the large greenhouse remained 1.5±0.3 to 7.4±2.1°C warmer during the daytime and 0.6±0.1 to 1.5±0.8°C warmer at night. The plant growth parameters were recorded higher in the large greenhouse. The mean marketable weight of cabbage cv. Golden Acre in the large greenhouse was significantly higher (619±53 g) than in the small greenhouse (523±121 g). Similarly, the marketable weight of cauliflower cv. Shantha was 599±35 g in the large greenhouse as against 537±42 g in the small greenhouse. Therefore, large passive solar greenhouses are recommended for farmers in Ladakh regions (Dolma et al., 2023b).

In view of our studies and the socio-economic condition of the farmers in the Ladakh region, we recommend that small greenhouses (9.8 m ×5.5 m × 2.4 m; L×W×H) are suitable for small farmers and those in remote areas.

Medium greenhouses (18.3 m × 7.3 m × 5.55 m; L×W×H) are best suited for large-scale adoption in Ladakh. Large greenhouses (27.4 m × 8.2 m × 2.7 m; L×W×H) are suitable for growing crops on a commercial scale.

8.3.4 Thermal blanket

Over seventy percent of the heat transfer occurs through the glazing material. The solar energy gained during day time is lost at night, especially from the areas covered under glazing material. Therefore, covering the glazing material with a thermal blanket after sunset significantly reduces heat loss at night. We found that covering the greenhouse with an insulation blanket (12 mm nitrile sheet with waterproofing slip) after sunset keeps the greenhouse warmer by 4°C at night in winter. However, since the greenhouse is small and most of the greenhouse produce is for self-consumption, using an insulating blanket is not practiced in Ladakh.

Many factors need to be considered while selecting a thermal insulation blanket for the greenhouse. The life of the material, its strength, initial cost, thermal properties, weight, and maintenance issues are the critical factors to consider. There are various insulation material options for greenhouse thermal blankets, and they all have advantages and disadvantages. The right material for thermal blankets in Ladakh condition depends primarily on insulation property, availability, and cost.

8.4 Types of passive solar greenhouses in Ladakh

Given the necessity of growing vegetables in winter, many passive solar greenhouses have been established in Ladakh since the 1980s. Some commonly known greenhouses are Ladakhi polyhouse, trench, polycarbonate, FRP, and tunnel. We developed the Ladakh Greenhouse, which is recommended as the most effective passive solar greenhouse for the Ladakh region. Polytrench has recently been developed, and a three year trial has been completed. The two widely used greenhouse designs for the Ladakh region are the traditional Ladakhi polyhouse and the recently developed Ladakh Greenhouse. The traditional Ladakhi polyhouse has been in use since the 1980s, while the Ladakh Greenhouse is being established on farmers' fields from 2020 onwards.

8.4.1 Ladakhi polyhouse

The Ladakhi polyhouse was developed in the 1980s by LEDeG. It is a popular greenhouse design due to its ease of construction and readily available construction materials. It has a mud wall on three sides, use polyethylene sheets as a cladding material, and has an angled wooden roof on the north wall of the greenhouses. The polyethylene sheet is removed from the greenhouse structure from June to October due to excessive heat that builds up inside the greenhouse. The increased use of the greenhouse has not only improved the dietary intake of vegetables during the winter months but also opened a new economic opportunity for the sale of early-season vegetables by the local farmers.

The temperature inside the Ladakhi polyhouse depends on the location and size of the greenhouse. The air temperature inside a Ladakhi polyhouse (19.8m length, 7.6m width, 2.4m height) remained below freezing in December and January. It remained 21.4±4.4°C warmer during the day and 9.6±1.6°C at night in winter (November to February). In summer (June to September), the air temperature inside the greenhouse remains 2.1±0.4°C warmer during the day and 0.3±0.2°C at night.

Two cycles of crops are grown each year in Ladakhi polyhouse. Leafy vegetables are grown in winter (mid-October to early March). From a single greenhouse, approximately 170 kg spinach is obtained. Vegetable nurseries are raised in spring (late March to early May), and 80000-100000 numbers of vegetable seedlings are obtained depending on the crop type. In summer, the polyethylene sheet is removed by over 96.2 percent of the farmers, and the remaining 3.8 percent partly removed the polyethylene sheet. Most farmers (over 91 percent) do not use the greenhouse during the summer months, and thus the structure remains underutilized (Angmo et al., 2019).

Farmers have been using the Ladakhi polyhouse for several reasons. The construction materials, except the polyethylene sheet, are locally available. Construction cost is low to moderate, depending on the size of the greenhouse. The earth walls on three sides and the floor store heat during the day and release it at night. Therefore, it remains much warmer than greenhouses with no heat retention walls (Angmo et al., 2019).

However, Ladakhi polyhouse has several limitations. The temperature inside the greenhouse often drops to sub-zero at night in December and January, and thus only freeze-tolerant leafy vegetables can be grown. The cladding material must be removed from the greenhouse structure from June to October due to excessive heat that builds up inside the greenhouse. The polyethylene sheet needs to be replaced after every 3–5 years. High wind speed, uneven surface of supporting frames, frequent removal, and extreme climatic conditions reduce the durability of the covering material. Rain and high humidity inside the greenhouse cause damage to the mud walls and the roof. Crops often get damaged due to the growth of molds and fungus on wooden structures. Therefore, the mud walls and roof need frequent repair and maintenance. The wooden door and ventilator frames are deformed due to the heavy load of the walls. The deformity leads to improper closing of the doors and ventilators, resulting in heat leaks from the gaps. Using polyethylene sheets as covering material is unsuitable for regions such as Changthang and Zanskar Valley, which receive heavy snowfall. The structure collapses due to the load of the snow.

8.4.2 Ladakh Greenhouse

We at DRDO-DIHAR designed and evaluated the Ladakh Greenhouse, which is an improvised naturally ventilated passive solar greenhouse, to overcome the limitations of the popular greenhouses used in Ladakh. The greenhouse was studied both during winter and summer for growing a variety of vegetables. A successful trial of the same led to making it in different sizes to cater to the needs of the farmers based on the availability of land and resources with them. The greenhouse is available in three different sizes: (1) Small greenhouses (9.8m×5.5m×2.4m; L×W×H), which is suitable for small farmers and those in remote areas; (2) Medium greenhouses (18.3m×7.3m×5.55m; L×W×H) is best suited for large scale adoption; and (3) Large greenhouse (27.4m×8.2m×2.7m; L×W×H) is best suited for large scale adoption. The greenhouse has several advantages over the previously used passive solar greenhouses. The first trial of the greenhouse on farmers' fields began in 2020.

Ladakh Greenhouse is significantly different from traditional greenhouses. The east, west, and north side walls are built with stone. The wall acts as a thermal mass that stores energy in the daytime and releases it during the night. Minimal maintenance is required after construction. The wall on the north side inside the greenhouse is used for fixing racks that can be used for different purposes. A transparent UV-stabilized 16 mm triple-layer polycarbonate panel is placed on the greenhouse's south-facing side. It has better heat retention capacity. The sheet retains good transparency even after 15 years of exposure to the harsh climatic conditions of Ladakh. The greenhouse has a sloped (to the north) coloured coated GI metal skin PUF (80-85 mm) roof on the north side that regulates solar radiation absorption on the inside surface. The GI does not allow the growth of molds and fungus, which is a major problem in the traditional greenhouse with a wooden roof. It is durable and adds to the aesthetics of the greenhouse. Rust-resistant pre-coated GI sheet topped with a layer of hay and soil as insulating material is also used for roofing. Manually operated ventilators with metallic frames on the south-facing frame and the west wall allow easy operation. The ventilation can be conveniently opened from inside the greenhouse without the requirement of any additional support.

The mean minimum temperature inside Ladakh Greenhouse remained above freezing during extreme periods in December and January. The night temperature increases by approximately 4°C when covered with a thermal blanket during winter. The greenhouse remains 24.1°C warmer during the day and 14.2°C at night in winter. In July, the greenhouse's maximum and minimum air temperature remained 42.1±4.0 and 16.1±1.3°C, respectively. Therefore, the greenhouse can be used even in peak summer months for growing warm-season crops.

Harvesting organic greenhouses tomatoes on 29 December 2022 inside a Ladakh Greenhouse

8.4.3 Advantages of Ladakh Greenhouse over traditional Ladakhi polyhouse

Ladakh Greenhouse is designed to overcome the limitations of traditional Ladakhi polyhouses. The mean temperature inside Ladakh Greenhouse remained above freezing points even in December and January in Leh. Therefore, plant growth is more vigorous in winter. Cabbage, cauliflower, broccoli, mushroom, and tomato are being grown successfully in peak winter in Ladakh Greenhouse, otherwise impossible in traditional greenhouses.

The polycarbonate sheet is fixed. Unlike polyethylene sheets in traditional greenhouses, there is no need to remove the cladding material even in summer months. The ventilation does not allow building excessive heat inside the greenhouse, even during summer months. The average service span of polycarbonate sheets is 15 years in Ladakh condition. It withstands high wind speed and extreme climatic conditions. Triple-layer polycarbonate as a cladding material is suitable for regions receiving heavy snowfall. The smooth curved design allows snow to slip off. The structure withstands the load of snow which remains on it.

The stone wall, metallic door, and ventilator frame require minimal maintenance. Racks are fixed on the north wall inside the greenhouse. The racks can be used for various purposes, thus using above-the-ground space in the greenhouse. The triple layer polycarbonate cover, fixed metallic structures, and the use of the wall as thermal mass result in higher heat retention due to high thermal retention properties and minimal heat escaping gaps.

The expected service life of Ladakh Greenhouse is 15 years. The walls, metallic frame, and PUF roof need minor maintenance to keep the greenhouse functional. Three cycles of crops are grown in Ladakh Greenhouse as against one or two crop cycles in the traditional greenhouses. The Ladakh Greenhouse has aesthetically pleasing interior space, which is important for an individual to form an attachment to the greenhouse. The major drawbacks of Ladakh Greenhouse are that the greenhouse structure is not easy to establish, and it requires qualified personnel to establish the structure. The material and installation costs are high.

8.4.4 Adoption of Ladakh Greenhouse at the grassroots level

DRDO-DIHAR formally transferred the Ladakh Greenhouse technology to the Agriculture Department for the large-scale establishment of the greenhouse on farmers' fields. The Agriculture Department of UT Ladakh is currently establishing Ladakh Greenhouse on farmers' fields in Leh and Kargil districts. Over 1500 Ladakh Greenhouses have been established in 120 villages within two years, and more greenhouse construction is underway. Farmers in Ladakh grow cauliflower, cabbage, broccoli, tomato, chili, and cucumber in peak winter months, otherwise impossible in the traditional passive solar greenhouses. Besides alleviating hidden hunger, a farmer earns Rs 60,000–80,000 annually from a single greenhouse. Due to its uniqueness and impact on the grassroots level, the

Ladakh Greenhouse technology received the Prime Minister Award for Excellence in Public Administration (under the Innovation category) on 21 April 2022. The UT Administration provides 75 percent financial assistance to farmers to establish Ladakh Greenhouse.

8.5 The way forward

Significant progress has been made in promoting naturally ventilated passive solar greenhouses. Many low- and medium-cost passive solar greenhouse structures have been designed and tested in Ladakh. The major focus has been growing leafy vegetables during winter and raising the nursery in spring. The Ladakh Greenhouse has been a technological leap, and farmers can grow various vegetables all year round. The prime focus to date has been on growing vegetables for household consumption. In the coming years focus should be on commercial and Hi-tech greenhouses. Diversification strategies must focus on promoting high-value crops with an expanding domestic and foreign market. Colored capsicum and cherry tomatoes, for instance, are in great demand in urban markets and fetch higher prices to the growers. Similarly, growing flower and ornamental crops instead of vegetables can increase farm income.

REFERENCES

Angmo P, Dolma T, Katiyar AK, Chaurasia OP, Stobdan T (2020). Growing cauliflower in winter under a passive solar greenhouse in trans-Himalayan Ladakh, India. *Def Life Sci J.*, 5: 192–197

Angmo P, Dolma T, Namgail D, Chaurasia OP, Stobdan T (2020). Growing cabbage (*Brassica oleracea* var *capitata* L.) in cold winter under a passive solar greenhouse in the trans-Himalayan Ladakh region. *Def Life Sci J.*, 5: 292–298

Angmo P, Dolma T, Namgail D, Tamchos T, Norbu T, Chaurasia OP, Stobdan T (2019). Passive solar greenhouse for round the year vegetable cultivation in trans-Himalayan Ladakh region, India. *Def Life Sci J.*, 4: 103–116

Angmo P, Dolma T, Phuntsog N, Chaurasia OP, Stobdan T (2022). Effect of shading and high-temperature amplitude on yield and phenolic contents of greenhouse capsicum (*Capsicum annuum* L.). *Open Access Res J Biol Phar.*, 4: 30–39.

Angmo P, Phuntsog N, Namgail D, Chaurasia OP, Stobdan T (2021). Effect of shading and high temperature amplitude in the greenhouse on growth, photosynthesis, yield and phenolic contents of tomato (*Lycopersicum esculentum* Mill.). *Physiol Mol Biol Plants*, 27: 1539–1546

Angmo, S., Angmo, P., Dolkar, D., Norbu, T., Paljor, E., Kumar, B. & Stobdan, T (2017). All year-round vegetable cultivation in trenches in cold arid trans-Himalayan Ladakh. *Def Life Sci J.*, 2: 54–58.

Dame J, Nüsser M (2011) Food security in high mountain regions: agricultural production and the impact of food subsidies in Ladakh, Northern India. *Food Sec.*, 3: 179–194.

Dar RA, Rather GM (2014) Assessment of the magnitude of malnutrition and related health problems in cold desert Ladakh-India. *Eur Acad Res.*, 2(4): 4895–4919

Dolma T, Kumar R, Namgail D, Chaurasia OP, Stobdan T (2023b) Size of passive solar greenhouse determines growth and yield of cauliflower and cabbage during winter in high mountain Ladakh region, India. *Def Life Sci J.*, 8: 137–142

Dolma T, Phuntsog N, Namgail D, Chaurasia OP, Stobdan T (2023a). Comparing heat unit requirements for flowering and fruit harvest of cucumber in an open field, shade net and greenhouse conditions. *Hortic Environ Biotechnol.*, 64: 345–353

Orzolek MA (2017). Guide to the Manufacture, Performance, and Potential of Plastics in Agriculture, 1st ed.; Elsevier: Cambridge, MA, USA, pp. 1–18.

Stobdan T, Angmo S, Angchok D, Paljor E, Dawa T, Tsetan T, Chaurasia OP (2018). Vegetable production scenario in trans-Himalayan Leh Ladakh region, India. *Def Life Sci J.*, 3: 85–92

Stobdan, T (2015). Plasticulture in cold arid horticulture. *DRDO Science Spectrum*, 155–159

Wittwer SH, Castilla N (1995). Protected cultivation of horticultural crops worldwide. *HortTechnology*, 5: 6–23

Chapter-9

Plasticulture: Use of Plastics in Crop Production in Ladakh

Plasticulture is a system of growing crops wherein a significant benefit is derived from using products derived from plastic polymers. Polyethylene polymer was discovered in the late 1930s. Its subsequent introduction in the early 1950s in the form of plastic films, mulches, and drip-irrigation tubing and tape, revolutionized the commercial production of selected vegetable crops and gave rise to plasticulture.

Plasticulture offers many benefits to the growers to realize greater returns per unit of land regardless of the size of an operation. The major benefits include earlier crop production, higher yields, higher quality produce, greater efficiency of water and fertilizer use, reduced leaching of fertilizer, reduced soil and wind erosion, a potential decrease in the incidence of diseases and insect pests, reduced competition with weeds, and reduced soil compaction. To realize these benefits, growers need to integrate the components of a plasticulture system (Lamont, 1996).

9.1 Early history

Greenhouse: The first plastic-covered greenhouse was erected in 1955 in England (Orzolek, 2017), and the use of plastics for greenhouse applications reached Ladakh in the late 1960s. An innovative low-cost greenhouse based on trench warfare was conceived in the late 1960s at DRDO-DIHAR, and vegetables were grown in trenches covered with polyethylene sheets (Angmo et al., 2017). The widespread use of plastic in greenhouses began in the 1980s when the Ladakh Ecological Development Group (LEDeG), a Ladakh-based volunteer organization, pioneered the construction of passive solar greenhouses with mud walls on three sides (north, east, and west) and polyethylene cover on the south-facing side (Rezvi, 2012).

Mulch: Plastic mulches have been used commercially on vegetables since the early 1960s (Lamont, 1996). It was not until 1990 the first field experiment on plastic mulch was conducted in Ladakh. Initial studies were conducted on cabbage, cauliflower, and tomato, and increased crop yield, early harvest, and water saving were recorded. Cucurbits such as bottle gourd, pumpkin, cucumber, and long melon were successfully grown in the open field using plastic mulch. However, despite the demonstration of the positive effect of mulch, the technology did not catch the attention of the farmers, extension personnel and administrators.

We conducted a survey in 2017 in Phey and Thiksey villages in the Leh district and found that 24.5 percent of the farmers had heard about the technology but had not used it before 2016 (Angmo, 2019). We revisited the potential use of the technology and conducted extensive studies from 2013–2018. One of my students took the potential use of the technology in enhancing vegetable productivity and crop diversification as a Ph.D. research topic. We then formally transferred the technical know-how to the State Agriculture Department on 12 Mar 2015, and mulching is included in the District Plan of Agriculture Department, Leh, from 2016–17 onwards. Black plastic mulch is now being used extensively by farmers in the region.

Low tunnel: Successful attempt to grow cabbage, knol khol, turnip, and leafy vegetables in winter under low tunnels was made in 1981.

9.2 Plasticulture: application in Ladakh

The two predominant uses of plastic film are as a cladding material in solar greenhouses and as a mulch film. Floating row covers have been used extensively for early nursery raising; however, their uses have declined in recent years. Low tunnel technology and drip irrigation systems are becoming popular with farmers. Polyethylene-lined water tanks are cost-effective, easy to establish, and effectively control seepage loss at a much lower cost. The use of colored shade nets is a relatively new concept in Ladakh, and its use is restricted to research farms. Plastic bags are used for nursery raising of cucurbits and extending the planting season of fruit crops. Plasticulture technology has a major role in post-harvest management, and plastic films are used for drying fruits and vegetables.

9.2.1 Plastic mulches

Mulching is a process of covering the soil's surface around the plants to create a congenial condition for growth. This helps in moisture conservation, temperature moderation, and weed control, accelerating plant growth and significantly increasing crop yield and quality. The materials used for laying over the soil may be organic plant residues, inert materials like pebbles, or synthetic materials like plastics. The mulch's color largely determines its energy-radiating behavior and its influence on the microclimate around a plant. Black plastic mulch is the predominant material used in Ladakh for growing vegetable and fruit crops. The plastic of 100-micron thickness is especially suitable for Ladakh's cool and dry conditions. It conserves moisture, increases soil temperature, and provides superior weed control. The technology is highly suitable for large-scale adoption by farmers in Ladakh, and it is the second most mass-adopted plasticulture technology in the region, next to the greenhouse. Although a variety of vegetables can be grown successfully using plastic mulches, tomato, capsicum, brinjal, muskmelon, watermelon, and cucumber have shown the most significant response.

Objectives of using plastic mulch in Ladakh

The overall objective of using black plastic mulch is to modify the soil environment around the plants to achieve optimal production of crops. The primary emphasis is enhancing yield, early harvest, quality produce, crop diversification, increasing water productivity, reducing weeds, increasing farm income, high return on investment, and reducing CO_2 emission.

Enhancing yields: Low temperature is one of the most important abiotic factors restricting the optimal production of warm-season vegetables. Black plastic mulch significantly increases the marketable yield of warm-season crops and makes growing selected warm-season fruits and vegetables possible on a large scale. The main reason for the higher yield is the increase in soil temperature. We found that the plastic mulch increases soil temperature by 4.9±0.3°C at the early growth stage of the tomato crop, which then reduces to 1.6±0.2°C at the last growth stage (Angmo et al., 2018a).

Our research has shown that the marketable yield of hybrid tomatoes increases by 107 percent while that of open-pollinated varieties by 80 percent when silver-on-black plastic mulch (30 microns) with the black surface facing the sun is applied. A marketable yield of 81.2±11.9 tonnes per hectare in hybrid tomatoes is achieved with mulch under a low input system (Angmo et al., 2018a). Using mulch increases the mean marketable yield of capsicum by 2.9-fold (Angmo et al., 2018b). The marketable yield of cucumber cv BSS-647 increased to 23.7±3.2 with mulch compared to 13.6±3.0 tonnes per hectare in non-mulch conditions. In the case of summer squash cv Australian Green, the marketable yield jumped from 52.6±8.4 under bare soil to 74.4±7.4 tonnes per hectare with plastic mulch (Angmo, 2019).

Black plastic mulch also increases the yield of root crops. The marketable yield of radish cv Pusa Himani increases by 10.7 percent in mulch condition (21.7±1.2) than in non-mulch (19.0±1.4 tonnes per hectare). In the case of beetroot, plastic mulch significantly increases the marketable yield by 21.9 percent. A marketable yield of 18.9 ± 0.7 is obtained with plastic mulch against 15.5 ± 1.1 tonnes per hectare in bare soil (Angmo, 2019).

Early harvest: The use of plastic mulch results in harvesting crops 1-2 weeks earlier as compared to traditional growing practices. Early crop harvest provides better marketing opportunities to the growers and fetches higher prices for their produce. Early harvest is defined as the first harvest during the season. Early fruit harvest of 11.3 ± 6.5 tonnes per hectare is obtained in hybrid tomatoes under plastic mulch compared to 3.5 ± 4.0 tonnes per hectare in a bare field. Similarly, an early harvest of 13.0 ± 11.7 tonnes per hectare in open-pollinated tomatoes is recorded against 3.5 ± 2.1 tonnes per hectare in bare fields (Angmo et al., 2018a). In the case of capsicum, cucumber, and summer squash, an early harvest of 0.8 ± 0.6, 1.2 ± 0.3, and 2.9 ± 0.5 tonnes per hectare, respectively, is recorded under mulch (Angmo et al., 2018b, Angmo, 2019).

Quality produce: Mulched plants are less prone to diseases and more uniform in size than those without. Mulching prevents fruits and flowers from being splashed by mud and water. It protects ripening crops such as tomatoes, pumpkins, and melons from direct contact with the soil.

Diversification towards high-value crops: Mulching allows farmers to grow high-value crops not traditionally grown in the open field. Watermelon, sun melon, and colored capsicum, for instance, grow well with mulch in the open field and maximize net profit for the growers.

Increasing agricultural water productivity: Water is a scarce resource in Ladakh. Due to the long cropping season and arid climatic conditions, 20 to 25 irrigations are required to raise vegetable crops in summer (Stobdan et al. 2018). We found that the use of mulching is an effective means of increasing water productivity. Black plastic mulch helps retain soil moisture by preventing water loss through evaporation. Water that condenses under the mulch drips back into the soil. Therefore, less water is required for growing crops. Approximately 120 liters of water are required to produce a kilogram of tomato in Ladakh conditions. But plastic mulching reduces the water requirement to just 50 liters. Farmers stated that the irrigation interval increased from 6.7 days to 11.8 days due to the use of mulch.

Reduces weeds: Labour cost is high in the Ladakh region. Unskilled labor charges Rs 700-800 daily to work on an agricultural farm. Therefore, engaging laborers for weeding on the vegetable farm is costly. The opaque surface of black plastic mulch effectively prevents sunlight from penetrating, so no weeds can grow underneath. Even if some weeds manage to poke through, they are easy to spot and pull out. Besides, fewer weeds present a reduced competition for water and nutrients with crops. Most farmers (51.8 percent) ranked weed suppression as the main benefit of mulching. The number of weeding reduced from 2.8 to 0.6 times on farmers' fields (Angmo, 2019).

Increasing farm income: Traditionally, cereals have been the staple crop of Ladakh. Wheat and barley occupy 61.8

percent of the total cropped area. A recent survey shows that income from cereals is Rs 57,000 per hectare, and there is scope to increase the farm income. Plastic mulch offers an excellent opportunity to diversify towards high-value crops. Shifting one-hectare land from staple crops to watermelon and sun melon can potentially increase gross returns. Farmers in the region have already started growing watermelon in the open field, earning Rs 10–12 lakh per hectare.

High return on investment: We worked out the return on investment of black polythene mulch in hybrid tomato production. A twenty-five-fold return is observed for every Rupee investment in purchasing and laying plastic mulch. Therefore, mulching technology offers a high return on investment and saves 50–60 percent water.

Easy ploughing: Small vegetable farms are generally ploughed manually. It is observed that the presence of higher soil moisture under plastic makes ploughing easier in spring as compared to bare soil.

Reduce CO_2 emission: Ladakh produces limited vegetables, partly because of low productivity. Therefore, most of the local market's fruit and vegetable requirements are   transported from outside the region. A survey that we conducted in 2015 concluded that approximately 21.4 ml of diesel is burned to transport one kilogram of fresh vegetables by truck from a nearby town at a 450 km distance (Angmo et al., 2018a). This leads to the emission of vehicular pollution in the fragile environment. Increasing crop productivity, early harvest, and crop diversification using plastic mulch will reduce carbon emissions and thereby contribute towards achieving the vision of Carbon Neutral Ladakh.

A farmer growing watermelon using mulching technology in Phey village

9.2.2 Solar greenhouse

The use of plastics for agriculture applications reached Ladakh in the late 1960s when the trench greenhouse was conceptualized. The widespread use of plastic in greenhouses began in the 1980s when the LEDeG pioneered the construction of a passive solar greenhouse with mud walls on three sides and polyethylene cover on the south-facing side (Rezvi, 2012). Since then, it has been widely used for greenhouse applications.

9.2.3 Drip irrigation

The use of drip irrigation systems is not widespread in the region. However, in recent years, the Agriculture

Department has promoted it for growing fruit crops and melons on barren lands. In Ladakh, special care is required for drip irrigation systems. Due to sub-zero temperatures in winter, pipeline joints must be unlocked every 50–60 feet in October to drain out the water entirely and prevent the pipeline system's breakage due to contraction and expansion. The pipelines are joined again in April to make the system operational for irrigation. The natural geographical gradient, common all over Ladakh, is often exploited to maintain the tank above the irrigation field level. The difference in the height of the tank and that of the irrigation field allow water to flow at the desired pressure without the need for electricity (Stobdan, 2015).

9.2.4 Coloured shade net

Crops respond to light intensity, quality, direction, and periodicity. Colored shade netting has been designed to manipulate plant development and growth. It provides physical protection, modify the micro-environment, increases the relative proportion of diffuse (scattered) light, and absorbs various spectral bands, thereby affecting light quality. These effects can influence crops and their associated organisms (Stamps, 2009). The photo-selective responses include fruit set, harvest time and yield, size, color, and internal and external quality.

Using colored shade nets in the horticultural crop is a relatively new concept in Ladakh. We conducted the first comprehensive study in 2013 and 2014. The effect of red shade net on tomato and capsicum was studied. Reduced evapotranspiration, reduced wind speed, increased relative humidity, and relief from excessive solar radiation have been recorded. The use of red shade net advances cucumber harvest by five days and increases yield by 39 percent (Dolma et al., 2023). An extensive field trial is required on diverse horticultural crops using different colored shade nettings.

9.2.5 Low tunnels

In low tunnels, plastic covers are supported above the crop by wire hoops to cover an individual row of vegetable crops in the field. The first successful attempt to use the low tunnel to grow winter vegetables was made in 1981–1982 at DRDO DIHAR. Cabbage, knol khol, turnip, spinach, mustard, and lye in the open field were covered with low tunnels in October–November. The tunnel was covered with a translucent plastic sheet, and at night an additional black plastic sheet was covered to minimize heat loss. Cabbage heads were formed in February, while leafy vegetables attained the harvestable stage in January. Peas started flowering in February, and the first harvest was in March.

Low tunnels find uses in spring for early crop harvest and in autumn to extend the harvest period. Low tunnel increases cabbage seed germination from 75.3 percent in open field to 91 percent. The seedling survival increased to 96.6 percent against 76.3 percent in the open field. Cabbage under a low tunnel reaches head maturity 13 days in advance without affecting the yield (Mehdi et al., 2014).

9.2.6 Walk-in tunnel

Walk-in tunnels are low-cost Quonset-shaped large structures with metal pipes framework covered with UV stabilized plastic sheet. This system is particularly appealing to new entry growers because of its low cost, simplicity, and effectiveness in protecting crops from low temperatures in spring and autumn. Too much temperature in the tunnel can be as much a problem as too little. These structures are used in spring for early vegetable harvest and in autumn to extend the harvesting period. The most significant use of the tunnels is for raising nursery seedlings in April-May. Since the tunnels do not have walls that act as thermal mass to store heat during the daytime, the use of tunnels for winter vegetable production is unsuitable. The tunnel temperature drops to -10°C or less at night in December and January.

9.2.7 Row cover

A row cover is a flexible, transparent covering installed over single or multiple rows of vegetables to enhance plant growth and yield. The cover may or may not be supported with hoops and is intended to be left over the plants for a relatively short period, usually 2–8 weeks, depending on the crop and weather conditions (Wells and Loy, 1985). Row covers are convenient to use and reduce installation costs. It is extensively used for nursery raising in the field in April-May. To increase the space between the germinating seeds and the polyethylene sheet, thin wires are tied on wooden sticks at four corners of the nursery bed, 2 feet above the ground level. The polyethylene sheet is removed after the seed germination during day time and covered back at night. The sheet is removed completely when the weather improves and the seedlings reach the transplanting stage. This helps increase seedling survivability when transplanted in the open field. Row covers have been used extensively for nursery raising when solar greenhouses were not many. However, their use has declined in recent years.

9.2.8 Soil solarization

Soil solarization with transparent polyethylene mulch film is a non-chemical approach to reduce weeds, pests, and diseases without using chemicals. It is suitable for commercial production in areas with high solar radiation. However, it is not being used in the region as a substitute for synthetic chemicals. In light of the recent initiatives on organic farming, soil solarization will likely become increasingly important. Around the world, solarization for disinfesting soil in open fields is being implemented at a relatively slow but increasing rate (Stapleton, 2000).

9.2.9 Polyethylene lined tank

Water is a scarce resource in Ladakh. Agricultural fields are irrigated mainly using water from glacier sources. Storage of water in large tanks and its subsequent controlled use for irrigation significantly reduces water wastage and ensures timely water availability for the crops. Polyethylene-lined tanks are cost-effective, easy to establish, and effectively control seepage. One such tank has been in use since 2008 at DRDO-DIHAR. It is made by digging the soil with an earth mover. After excavation, the side slopes are made smooth and compact by removing angular projections, protruding stones, pebbles, roots, etc. Before laying the sheet, a thin slurry of clay and cow dung is applied on the walls to avoid damage by sharp objects. The edge of the polythene is anchored inside the soil and weighed down with sandbags. The tanks are lined with 300–500 GSM UV-stabilized polythene sheets. The expected life of polyethylene is ten years. However, fencing the tank is a must to prevent accidental drowning of children and stray animals. Significant saving in water has been observed when such tanks are integrated with a drip irrigation system (Stobdan, 2015). However, its use is not popular in the region.

9.2.10 Nursery bag

Low-density polyethylene bags, usually black, are used for cucurbit seedling production and extending the planting season of young fruit trees. Raising vegetable seedlings in nursery bags has recently been replaced by direct seeding under plastic mulch without affecting the yield. Plastic bags are now being used to transplant fruit saplings so that the young trees can be transplanted in the field anytime in the year, except in winter.

9.2.11 Sun drying

Sun drying is the most economical and ecologically sound method, but it is slow, weather-dependent, and exposes the product to food safety hazards. Low-cost solar dryer, using plastic sheets, accelerates drying and produces a hygienic product. Such dryers are extensively used to dry surplus fruits and vegetables, especially apricots.

9.3 The way forward

Significant progress has been made in promoting plasticulture in Ladakh. In the coming years focus should be on the use of biodegradable film, photo-selective film, and colored mulch. The combined use of mulch and drip irrigation in a greenhouse must be studied. Focus attention is needed in using technology for diversification towards high-value crops and managing insect pests. Studies related to the long-term influence of film mulching on the soil microbial community are crucial.

Biodegradable mulch film: The disposal of agricultural plastic is a major problem. The polyethylene plastic mulch film is not biodegradable and cannot be ploughed back into the soil. It undergoes fragmentation, and small fragments are blown over and find their way into other pristine environments. Burning the plastic film after its intended use is not an option. Therefore, the conventional black plastic sheet must be replaced with biodegradable plastic mulch. Unfortunately, 'biodegradability' is a much-misused term, and many misleading claims abound in the marketplace. Biodegradability is an end-of-life option that harnesses the power of microorganisms present in the selected disposal environment via the microbial food chain (Narayan, 2017). Therefore, the market's biodegradable plastics must be analyzed carefully before any recommendation is made for large-scale adoption. Biodegradation depends on the environmental condition and intended use; therefore, one must identify the disposal environment.

Photo-selective film: Farmers in the region exclusively use the traditional black mulch film, which blocks solar radiation and heats the soil. Recently, new types of photo-selective films with different colors have been available. Contrarily to the traditional mulch film, the photo-selective film blocks only the PAR of solar radiation by modulating the part useful to heat or cool the soil. These films have optical properties, such as selecting the incoming solar radiation and outgoing radiation emitted from the ground. These films have significantly more soil heating or cooling effect, greatly contribute to water saving, and support pest management (Mormile et al., 2017). There is a need to undertake studies on selecting photo-selective mulch, which is superior to the traditional black mulch film in terms of higher crop yield, shortened time to harvest, water saving, and reduced plant pests.

Colored mulch: Coloured polyethylene mulches reflect different red and far-red light radiation patterns into a crop canopy. These consequent effects on photosynthetic efficiency and plant morphogenesis may increase yields (Decoteau, 1989). Production of many fruits, roots, tubers, and bulbs have been recorded using colored plastic mulches. Further research is needed to benefit from using colored mulch compared to traditional black plastic.

Combined use of mulch and drip irrigation system: Water is a scarce resource in the Ladakh region, and mulch and drip irrigation systems greatly contribute to saving water. The use of mulch is becoming popular among farmers. However, the drip irrigation system is a relatively new technology in the region. Combining mulch with drip irrigation is necessary to maximize water saving. Studies need to be undertaken to evaluate the combined use of drip irrigation with mulching for fruit and vegetable cultivation in terms of yield, water use efficiency, and economics.

Mulching in the greenhouse: Frequent irrigation requirement is a major problem in using naturally ventilated greenhouses in summer. The crop under the greenhouse must be irrigated twice a week against weekly intervals in the open field. The use of mulch can effectively reduce the water requirement. Using photo-selective mulch for soil heating during cold seasons may increase crop yield. Further research is needed to benefit from using photo-selective film in the greenhouse.

Diversification towards high-value crops: Mulching technology has opened new avenues for the farmers in the region. Diversification strategies must focus on promoting high-value crops with an expanding domestic and foreign market. Red and yellow capsicum, for instance, are in greater demand in urban markers than the traditional green capsicum. Similarly, growing flower and ornamental crops instead of vegetables can increase farm income.

Management of insect pests: Aphids are the most common insect that causes serious damage to fruit and vegetable crops in the region. Crops infested with the aphid cause a 41 percent reduction in cauliflower yield and a 35 percent yield reduction in knol-khol and radish under protected cultivation (Singh and Dhiman, 2018). The use of silver reflective mulch is an effective method in the management of aphids. Consistently lower aphid numbers on plants growing over mulches than those growing over bare soil has been reported. Therefore, using reflective mulch to manage aphids and other major insect pests must be studied.

Influence of film mulching on soil microbial community: The soil microbial community plays a crucial role in nutrient cycling, and it is a sensitive indicator of soil quality. Just as different groups of microorganisms vary in their ability to adapt to the various environmental conditions, changes in soil microbial community under plastic mulch application are expected. Information about soil microbial communities and the effect of plastic film mulching in cold arid regions remains scarce. Therefore, research on the effects of plastic films on soil microbial communities has become a fundamental aspect of sustainable agriculture.

Disposal of used plastic: Disposal of used plastic is a major concern. Currently, there is no mechanism for the disposal of the used film. Generally, used plastic is recycled or properly incinerated to capture the energy locked inside these products. The use of biodegradable film could be an answer to solve the problem.

REFERENCES

Angmo S (2019). Crop diversification and vegetable productivity enhancement using black polyethylene mulch in trans-Himalayan Ladakh, India. PhD thesis submitted to the Bharathiar University.

Angmo S, Bhatt RP, Paljor E, Dolkar P, Kumar B, Chaurasia OP, Stobdan T (2018a). Black polyethylene mulch doubled tomato yield in a low-input system in the arid trans-Himalayan Ladakh region. *Def Life Sci J.*, 3: 80–84

Angmo S, Dolkar D, Norbu T, Kumar B, Stobdan T (2018b). Black polyethylene mulch results in over two-fold increase in capsicum (*Capsicum annuum* L.) yield in trans-Himalaya. *Natl Acad Sci Lett.*, 41: 173–176

Angmo S, Angmo P, Dolkar D, Norbu T, Paljor E, Kumar B, Stobdan T (2017). All year round vegetable cultivation in trenches in cold arid trans-Himalayan Ladakh. *Def Life Sci J.*, 2: 54–58

Decoteau DR (1989). Mulch surface color affects yield of fresh-market tomatoes. *J Amer Soc Hort Sci.*, 114: 216–219

Dolma T, Phuntsog N, Namgail D, Chaurasia OP, Stobdan T (2023). Comparing heat unit requirements for flowering and fruit harvest of cucumber in open field, shade net and greenhouse conditions. *Hortic Environ Biotechnol.*, 64: 345–353

Lamont WJ (1996). What are the components of a plasticulture vegetable system? *HortTechnology*, 6(3): 150–154

Mehdi M, Ahmad F, Hussain N (2014). Low tunnel technology is a valuable vegetable forcing intervention for boosting economy of farmers in Ladakh. *Asian J Hortic.*, 9: 106 –108

Mormile P, Stahl N, Malinconico M (2017). The World of Plasticulture. In: Malinconico, M. (eds) *Soil Degradable Bioplastics for a Sustainable Modern Agriculture. Green Chemistry and Sustainable Technology.* Springer, Berlin, Heidelberg

Narayan R (2017). Biodegradable and Biobased Plastics: An Overview. In: Malinconico, M. (eds) *Soil Degradable Bioplastics for a Sustainable Modern Agriculture. Green Chemistry and Sustainable Technology.* Springer, Berlin, Heidelberg

Orzolek MA (2017). *Guide to the Manufacture, Performance, and Potential of Plastics in Agriculture*, 1st ed.; Elsevier: Cambridge, MA, USA, pp. 1–18

Rezvi J (2012) *Ladakh: crossroads of high Asia.* 3rd Edition, Oxford University Press, New Delhi, India

Singh N, Dhiman S (2018). Quality and quantity loss by aphid infestation in vegetables grown under protected cultivation in Ladakh region. *Def Life Sci J.*, 3: 71–74

Stamps RH (2009) Use of colored shade netting in horticulture. *HortScience*, 44: 239-241

Stapleton JJ (2000). Soil solarization in various agricultural production systems. *Crop Prot.*, 19: 837–841

Stobdan T, Angmo S, Angchok D, Paljor E, Dawa T, Tsetan T, Chaurasia OP (2018). Vegetable production scenario in trans-Himalayan Leh Ladakh region, India. *Def Life Sci J.*, 3: 85–92

Stobdan T (2015). Plasticulture in cold arid horticulture. *DRDO Science Spectrum*, 155–159

Wells OS, Loy JB (1985). Intensive vegetable production with row covers. *HortScience*, 20(5): 822–826

Use of black plastic mulch for growing cherries

Chapter-10

Emerging Insect Pests and Diseases in Ladakh

Ladakh region is known for its low incidence of insect pests and diseases, which can be partly attributed to low humidity, freezing winters, low cropping intensity, and diversified cropping system. Moorcroft, who stayed in Ladakh for two years, from September 1820 to September 1822, stated in his travelogue that apple trees are "*very rarely attacked by disease.*" He mentioned the wheat crop and said, "*There is very rarely any disease amongst the crops, although after a fall of rain heavier than ordinary, a few ears may sometimes be affected with ergot, or speckled with mildew*" (Moorcroft and Trebeck, 1837).

The incidence of disease and insect-pest attack is becoming a serious problem in recent years. Due to the low incidence of insect pests and diseases in earlier times, farmers' awareness of the problem is low. We conducted a survey in 2021 among 500 apple growers in the Leh district and found that over 90 percent of the respondents could not mention the name of a single insect pest or disease of apples. However, the codling moth (*Cydia pomonella* L.) in apple and onion maggots (*Delia antique*) is becoming a serious pest causing substantial economic loss. Yellow tail moth (*Euproctis similis*), which remained unseen in the region, has cracked havoc on apricot trees during 2013–16. Aphid is a regular pest of vegetables and apricot in the region. Migratory locust (*Locusta migratoria* L.) emerged for the first time as a severe pest in Zanskar and Changthang Valley in August 2006. Overall, with the incidence of few established insect pests and diseases, and sudden outbreaks of previously unknown insects, it is essential to be prepared for a potential outbreak of insect pests in the region.

10.1 Wheat

The major diseases of wheat in Ladakh, in descending order, based on disease severity, are yellow rust (*Puccinia striiformis* Westend), molya (*Heterodera avenae*/ cereal cyst nematode), foot/root rot (*Rhizoctonia solani* Kuhn), ear Cockle, tundu, loose smut (*Ustilago segatum* var. *tritici*), foliar blight and powdery mildew (*Blumeria graminis* f. sp. *tritici*). In severely infected crops, the grain yield reduces by 62 percent in yellow rust, 13 percent in powdery mildew, 18 percent in foliar blight, 53 percent in molya, and 44 percent in foot/ root rot. Seed treatment, management of irrigation water, crop rotation and use of resistant varieties are recommended to manage the diseases. Zinc deficiency is a major problem, and 12 percent of the wheat crop in the region shows a prevalence of zinc deficiency (Vaish et al., 2013).

Stripe (yellow) (*P. striiformis* Westend) and stem (black) (*Puccinia graminis* Pers. f. sp. *tritici* Eriks & Hans) rusts are common wheat diseases in Ladakh. However, leaf (brown) (*P. triticina* Eriks) rust does not occur much, and its incidence is restricted to a few pustules only (Bhardwaj et al., 2012). Yellow rust causes a 62 percent reduction in grain yield in severely infected crops, while in moderately and slightly infected crops, grain yield reduces by 13 and 5 percent, respectively (Vaish et al., 2013). The high incidence of rust is because of the extensive use of landraces susceptible to rust. The present-day wheat varieties released for northern India are resistant to *Puccinia graminis* and *P. striiformis*. Therefore, replacing local varieties with rust-resistant and improved wheat varieties is recommended to reduce the incidence of rust. The regionally suited Sinchen variety is resistant to wheat rust (Bhardwaj et al., 2012).

10.2 Barley

The major diseases of barley in Ladakh are yellow/ stripe rust (*Puccinia striiformis* f. sp. *hordei*), powdery mildew (*Blumeria graminis* f. sp. *hordei*), leaf spot blotch/ blight (*Cochliobolus sativus*, *Altenaria alternata*), covered smut (*Ustilago hordei*), loose smut (*U. nuda* var. *hordei*), foot/ root rot (*Rhizoctinia solani*, *C. sativus*) and cereal cyst nematode (*Heterodera avenae*). In severely infected crops, the grain yield reduces by 66 percent in yellow rust, 63 percent in foot/root rot, 58 percent in cereal cyst nematode, 53 percent in leaf spot blotch/blight, 37 percent in powdery mildew (Vaish et al. 2011). Seed treatment, management of irrigation water, crop rotation, and use of resistant varieties are recommended to manage diseases. Zinc deficiency is a major health problem, and barley crops in the region show the prevalence of zinc deficiency in 14 percent of the crops (Vaish et al., 2011).

10.3 Apple

The codling moth (*Cydia pomonella* L.) is a serious insect pest of apples in Ladakh. The pest is widely distributed in fruit-growing areas of Ladakh. It is the most persistent, destructive, and difficult to control pests of fruit crops in Ladakh. It causes two types of fruit damage: stings and deep entries. Stings are entries where larvae bore into the flesh a short distance. Deep entries occur when larvae penetrate the fruit skin, bore to the core, and feed in the seed cavity. Larvae may enter through the fruit's sides, stem, or calyx ends. One or more holes plugged with frass on the fruit's surface are a characteristic sign of codling moth infestation. Calyx entries are difficult to detect without cutting the fruit. The pest is reported to infest 42.2-76.8 percent of the apple trees in the region (Hussain et al., 2015). To avoid and restrain its spread to other parts of the country, the Government has imposed restrictions under SRO 47 dated 25th February 2008 under the Jammu and Kashmir Plant Diseases and Pests Act 1973 on the movement of plants and fruit of apple, apricot, and walnut from Ladakh. The pest completes two-and-a-half generations a year. The overwintering larvae pupate in April. The first-generation adults appear in May and then lay eggs after mating. Newly hatched larvae of the pest bore a short distance into the flesh and continued feeding, boring deep into the center to the seeds, and ultimately infested fruits fell. They exit the fruit and pupate under the bark or debris under the tree. The second-generation adults appear during the last week of June to the first week of July. The pest passes the winter as full-grown larvae diapause inside a thick silken cocoon, mainly under the tree's bark. It remains dormant throughout the winter and withstands the harsh winter months (Ali et al., 2020).

The Woolly apple aphid, *Eriosoma lanigerum* (Hausmann), is a pest of apples worldwide. It gets its name from the woolly appearance of its colonies. The insect is a sucking pest that weakens the tree by feeding on limbs and roots. It feeds on new terminal leaves, causing them to curl or form rosettes. Heavy infestation may cause deformed twigs and branches, which can be a stressed tree. It is the feeding on the roots that produce the most significant damage. Mature trees usually suffer minor damage from root infestations, but root infestations damage young trees (Bessin, 2017). Woolly aphid is not a serious pest of apples in Ladakh, but its incidence has been observed in a few places. Adult aphids are approximately 1.5 mm long and wingless in summer. They are reddish-purple but usually entirely covered by a white waxy material. Eggs, overwintering in bark cracks, hatch in spring, and nymphs feed on new growth. Later in the season, a winged generation develops, which migrates and lays eggs. There are several generations in a year. Adult females over-wintering on the roots continue to develop and reproduce slowly. Spring migration of crawlers from the roots into the tree is the major source of re-infestation each year. Treatments of woolly apple aphids are recommended when 10 percent of the pruning scars are infested with live colonies (Bessin, 2017). Rootstocks vary in susceptibility to woolly apple aphids, and susceptible rootstocks form galls around the infestation sites. It is recommended to use the Malling-Merton series of rootstocks (e.g., MM.106 and MM.111) resistant to woolly apple aphids.

10.4 Apricot

Yellow tail moth (*Euproctis similis*) is not a regular insect pest of fruit crops in Ladakh. However, it emerged as a major pest of apricot from 2013–2016 in the Dha-Hanu and Sham belt of the Leh district. The caterpillar causes extensive defoliation to host trees through feeding activities, leading to reduced growth and destruction of fruits. The infestation of the insect caused a severe loss of over 80 percent in income for the villagers in the Dah-Hanu belt during 2013–16. Besides apricot, the insect was also found feeding on apple, peach, rose, poplar, and willow. The hair on the caterpillar causes skin rashes and respiratory problems among the villagers. Local inhabitants perceived the severe infestation as a result of climate change and disrespect for local deities. If appropriate control measures were not taken, the pest could destabilize the entire apricot industry of the region. The pest has been successfully managed in the Leh district, and no incidence has been observed since 2019 (Stobdan et al., 2019).

Aphid is a regular pest of apricot in Ladakh. The pest is not known to cause extensive damage to apricots in the region. However, in 2016 there was an outbreak of the pest, causing a substantial economic loss to Leh district's major apricot growing belt, particularly in the Sham and Turtuk belt. Aphid infestation on apricot in 2016 alone resulted in a loss of Rs 545 lakhs in the Leh district. The pest infestation was particularly severe on *Halman* cultivars compared to others. It damages floral and vegetative buds by sucking sap, due to which unfolding leaves curl up, remain stunted, get distorted, and later turn pale. Consequently, the fruit set is poor with a premature fruit drop and forming of sub-normal fruit. Severely infected ones get their leaves curling inwards and eventually dry out and shed.

Gummosis is a major problem of stone fruits in cold desert conditions. Exudation of gum from the stem and branch loosens the bark causing secondary infection and leading to the drying of branches in severe cases. The causal factor could be biotic such as fungi and bacteria (*Pseudomonas syringae*), or it may be due to abiotic factors. To manage the disease, scarifying the gumming area with a sharp knife in spring, followed by applying Mashobra paste, is effective. A foliar spray of streptomycin (0.01 percent) plus copper oxychloride (0.2 percent) is effective in overcoming gummosis (Dwivedi et al., 2007).

Yellow tail moth infestation- apricot tree without a single leaf in summer

10.5 Vegetable crops

The major insect pests found damaging different vegetable crops in the region are *Helicoverpa armigera* Hüb, *Pieris brassicae* Linn., *Plutella xylostella* Linn., *Phyllotreta* spp., *Agrotis ipsilin* Hufn., *Etiella zinckenella* Treitschke., *Thrips tabaci* Lind., *Hylemya antique* Meigen., *Adoretus* spp., *Tetranychus urticae* Koch. (Pandey et al., 2006).

Cabbage aphids (*Brevicoryne brassicae* Linn.) are the most common aphid species that cause serious damage to cabbage, cauliflower, broccoli, and Brussels sprouts. It also attacks Chinese cabbage, radish, and kale. The aphids are gray-green but usually appear gray or white due to a dusty, waxy secretion that covers their bodies. Aphids feed by sucking sap from their host plants. Cabbage aphids prefer to feed on young leaves, flower buds, or seed stalks in the upper part of the plant. It clusters underneath the leaves and on growing points. Feeding injury includes wrinkled and downward-curling leaves, yellow leaves, and reduced growth. For effective pest management, the fields should be scouted weekly for signs of aphids from spring onwards. Quick action should be taken before a damaging population develops, especially on more vulnerable young plants. On older established plants, the aphid can usually be tolerated. Use finger and thumb to squash aphid colonies where practical. Clip off badly infested leaves and immediately dispose of them in an active compost pile. The field should be ploughed immediately after harvest to prevent the spread of aphids to other crops. Destruction of old

Brassica plant debris at the end of the season can reduce the risk of the next season's plants becoming affected.

Onion maggot (*Delia antique*) is a serious pest in major onion-growing areas in Ladakh. The pest results in 50–65 percent onion bulb yield loss in severely infested villages (Gupta et al., 2021). Onion maggot overwinters as pupae in the soil associated with culled onions in the field or onion cull piles. Adults emerge around mid-May and mate over three days, after which they begin laying tiny white eggs at the base of the host plant. The larvae, upon emergence, crawl beneath the leaf sheath and enter the bulb. The maggot pupates in the soil, and a second generation of adults appears three to four weeks later. The pest completes three generations yearly, but the first generation is the most damaging. The third generation attacks onion bulbs in mid-August, shortly before harvest, which leads to storage rot (Delahaut, 2001).

Management of the maggot using insecticide and biopesticide is ineffective. Studies conducted in Ladakh showed 13.6 to 18.6 percent incidence of onion maggot when treated with insecticides as against 30.8 percent incidence in untreated soil (Gupta et al., 2021). Development of insect resistance against insecticides is reported. Hence, the use of pesticides is not a viable option when farmers and extension workers are unaware of the judicious use of pesticides. Managing the onion maggot through cultural practices is a viable option for the region. High mountain geographical barriers, isolation distance between the villages, and crop rotation can significantly reduce the maggot population. Onion maggots are highly host-specific to plants in the onion family, including onions, leeks, shallots, garlic, and chives. If the entire farming community in a village decides not to grow onion and related species for one season, the chances of re-infestation will reduce significantly. The chances of the spread of the insect from nearby villages are minimal because of the high mountain geographical barrier and long isolation distance. If the incidence is low, the overwintering populations of onion maggots can be reduced by destroying crop debris and removing culls from the field. Onion sets can also be planted a few weeks before fly emergence is predicted. Farmers should strictly avoid procuring onion seedlings from maggot-infested farms and villages. Onion varietal selection is important to reduce the loss due to onion maggot. The incidence of onion maggot in the Red Coral variety is 11 percent, as against 27 percent in Pinnari (Ganie et al., 2019). Therefore, growing less susceptible high-yielding varieties is recommended. Crop rotation with non-Allium crops results in a 32 percent reduction in yield loss (Gupta et al., 2019). Therefore, an integrated approach combining crop rotation, varietal selection, soil solarization, geographical barrier, and use of biopesticides and onion sets needs to be tested to manage the severe pest of onion in the region.

The major diseases and pests of potatoes in the region are early blight (*Altenaria solani*), late blight (*Phytophthora* spp), fusarium dry rot, fusarium wilt (*Fusarium* spp), leaf spot and necrosis (*Cochliobolus* spp). The incidence of the diseases is higher in villages located at a lower altitude than in higher altitude areas. Similarly, the disease incidence is higher in places with higher atmospheric relative humidity than that with lower humidity (Phour et al., 2018).

10.6 Greenhouse vegetables

Cabbage aphids (*Brevicoryne brassicae*) are the most common aphid species that cause serious damage to crops under passive solar greenhouses. Infestation is seen in summer as well as in the winter season. Crops infested with the aphid species cause a 41 percent reduction in cauliflower yield and a 35 percent yield reduction in knol-khol and radish under protected cultivation (Singh and Dhiman, 2018).

The Diamondback moth, *Plutella xylostella* L., is a serious pest of crucifer crops worldwide. However, it is not a severe pest in the Ladakh region. Its incidence is observed in greenhouses. The diamondback moth's short generation time and high fecundity may make it a significant pest of crucifer crops in this region. Though insecticides remain the primary defense against the diamondback moth, insecticide resistance has become a problem in many diamondback moth populations. As a result, alternative insect control methods, including plant-based resistance, need to be investigated.

Cutworm is a serious pest of vegetables in the greenhouse and open-field in the region. The larvae come in various colors and patterns but always appear smooth-skinned to the naked eye. They coil up into a little ring when disturbed. They hide in the soil during the day and emerge at night to feed on the young plant. Early in the season, the cutworms cause stand loss by cutting off seedlings at the soil line. Cutworm incidence is often associated with the residue of host plants remaining in the field before planting and surrounding weedy plant matter. As most cutworm species have a broad host range, tillage at least two weeks before planting helps destroy plant residue that could harbor larvae and pupae. Detecting the cutworms early helps to control them and avoid serious damage. Ploughing helps expose the cutworm to natural predators and sunlight, eventually killing them. Small infestations might be controlled by digging out the damaged seedling to find and kill the cutworm. Cultural control, *Bacillus thuringiensis*, and the Entrust formulation of spinosad are organically acceptable management tools. Treatment should be done when the presence of cutworms is detected. Cutworms are usually localized within a field, so consider marking the areas where damage is observed and treating only those areas (Natwick et al., 2013).

10.7 Forest plantation

Poplar and willow are the major forest plantation of the region. The major insect pests of poplar and willow in the region are gypsy moth (*Lymantria* spp), willow scale (*Chionaspis salicis* L.), tent caterpillar (*Malocosoma* spp.), willow leaf beetle (*Altica* spp.), goat moth (*Cossus cossus* L.), ermine moth (*Yponomeuta rorellus* Hubner), poplar petiole gall insect (*Pemphigus* spp.), willow apple gall (*Pontania* spp.). Over 60 percent of willow plantations in the Leh district are infested with scale insects, while willow apple gall inducers are found to cause severe damage in the Zanskar block in Kargil, and Nubra and Khalsi block of Leh district (Pandey et al., 2007). Poplar leaf minor (*Phyllonorycter populifoliella*) is a major insect pest of poplar in Ladakh. It is highly abundant in the region but has not been reported from any other region in India. The tree's lower branches are highly infested, carrying up to nine mines per leaf. Noticeable damage is seen in July-August. The moth seems to develop in two generations — the mines of the first generation in May and the mines of the second generation in late July till late August (Shashank et al., 2021).

10.8 Migratory locust

The migratory locust (*Locusta migratoria* L.) is a polyphagous insect pest that damage a wide range of agricultural, horticultural, and forestry plantation. The locust emerged for the first time as a serious pest in Zanskar and Changthang Valley in August 2006. The infestation was found both in cropped and pasture areas. No crop loss has been reported in Changthang Valley, but 90 percent of crop damage has been reported in Zanskar Valley. The outbreak inflicted serious loss to the pashmina wool industry and animal husbandry by destroying important pasture land. The locust infestation was found along the sides of river Indus from Demchok to Kiari in Changthang, whereas the locust infestation in Zanskar valley was reported in a 1500 ha cropped area beside the pasture land. The Locust Warming Organization (LWO) of the Directorate of Plant Protection, Quarantine and Storage deployed 30 micro ULVA, and ten power sprayer teams to undertake control operations. A 300 ha area in Changthang and 1000 ha in Zanskar Valley were treated using 1300 liters of Chloropyrifos 20% E.C. (Asre et al., 2005). The outbreak of its small swarms occurred in 2007 and 2008 (Veer et al., 2013).

REFERENCES

Ali L, Prasanth CS, Ali I, Fatima N (2020). Invasive pest of horticultural crops in Jammu & Kashmir and Ladakh region. *Int J Recent Sci Res.*, 11: 38316–38322

Asre R, Singh S, Zadoo RK, Chandurkar PS (2005). Outbreak of migratory locust: *Locusta migratoria* L, in Ladakh region of Jammu and Kashmir State during 2006. *Plant Prot. Bull.*, (Faridabad) 57 (3/4): 13–19.

Bessin R (2017). Woolly apple aphid. Entfact-219. University of Kentucky College of Agriculture, Food and Environment, Lexington, KY 40546

Bhardwaj SC, Gangwar OP, Singh MS, Saharan MS, Sharma S (2012). Rust situation and pathotypes of *Puccinia* species in Leh Ladakh in relation to recurrence of wheat rusts in India. *Indian Phytopath.*, 65: 230–232.

Delahaut KA (2001). Onion maggot. The University of Wisconsin Extension Bulletin A3422

Dwivedi SK, Kareem A, Ahmed Z (2007). Apricot in Ladakh. Field Research Laboratory, Leh Ladakh, India. 47 p.

Ganie SA, Wani BA, Wani MA, Zargar BA, Mir NA, Safal R (2019). Evaluation of different onion varieties for morphological traits, yield and maggot incidence under cold arid conditions of Ladakh. *J Entomol Zool Stud.*, 7(3): 202–205

Gupta V, Namgyal D, Kumar A, Namgyal D, Angchuk S, Safal R (2019). Assessment of integrated pest management against onion maggot in trans-Himalayan Leh. *Int J Curr Microbiol Appl Sci.*, 8(10): 2180–2183

Gupta V, Raghuvanshi MS, Namgyal D, Landol S, Dorjey S, Stanzin J, Tundup P (2021). Bio-efficacy of different insecticides/ bio-insecticides against onion maggot (*Delia antique*) under in-vitro conditions in cold arid region of India. *Pharm Innova J.*, 10: 1254–1258

Hussain B, Ahmad B, Bilal S (2015). Monitoring and mass trapping of the codling moth, *Cydia pomonella*, by the use of pheromone baited traps in Kargil, Ladakh, India. *Int J Fruit Sci.*, 15: 1–9

Moorcroft W, Trebeck G (1837) Travels in the Himalayan provinces of Hindustan and the Panjab in Ladakh and Kashmir; in Peshawar, Kabul, Kundiz, and Bokhara. Gyan Publishing House, New Delhi-110002

Natwick ET, Stoddard CS, Zalom FG, Trumble JT, Miyao G, Stapleton JJ (2013). UC IPM Pest Management Guidelines: Tomato, UC ANR Publication 3470

Pandey AK, Namgyal D, Mehdi M, Mir MS, Bilal AS (2006). A case study: major insect pest associated with different vegetable crops in cold arid region Ladakh, of Jammu and Kashmir. *J Entomol Res.*, 30: 169–174.

Pandey AK, Namgyal D, Mir MS, Bilal AS (2007). Major insect pest associated with forest plantations in a cold arid region, Ladakh of Jammu and Kashmir. *J Entomol Res.,* 31: 155–162

Phour M, Singh N, Ghai A, Tiga S, Rinchen T (2018). Current health status of potato crop in different altitude regions of Ladakh, Jammu and Kashmir, India. *Int J Agri Environ Biotechnol.,* 11(2): 303–309

Shashank PR, Singh N, Harshana A, Sinha T, Kirichenko N (2021). First report of the poplar leaf miner, *Phyllonorycter populifoliella* (Treitschke) (Lepidoptera: Gracillariidae) from India. *Zootaxa,* 4915 (3): 435–450

Singh N, Dhiman S (2018). Quality and quantity loss by aphid infestation in vegetables grown under protected cultivation in Ladakh region. *Def Life Sci J.,* 3: 71–74

Stobdan T, Deen M, Gupta V and Raghuvanshi MS (2019). Integrated management of yellow tail moth (*Euproctis similis*) on apricot in Leh Ladakh India. LAHDC Leh, Ladakh

Vaish SS, Ahmed SB, Prakash K (2011). Documentation of wheat diseases of the trans-Himalayan Ladakh region of India. *Crop Prot,* 30: 1129–1137

Vaish SS, Bilal S, Ahmed, Prakash K (2013). Documentation of wheat diseases of the trans-Himalayan Ladakh region of India. *J Wheat Res.,* 4(2): 22–23

Veer V, Sharma AK, Tikar SN, Mendki MJ, Tyagi V, Chandel K, Selwamurthy W (2013). Molecular characterization of migratory locust, Locusta migratoria Linn. (Orthoptera: Acrididae: Oedipodinae) from Ladakh region, India. *Int J Vet Med: Res Report*, article ID 942894

Chapter-11

Key Challenges Facing Agriculture in Ladakh

The land-based economy of Ladakh is under siege today. With more and more employment avenues open to the locals in the form of the army, tourism, and sundry jobs, their need to depend on the land-based economy is decreasing. However, the euphoria could prove to be a short-lived one due to several factors. The current economic boom that Ladakh is experiencing cannot be considered sustainable. As a result, the attitude of disregard towards the land-based economy that most Ladakhis nowadays sport could cost them dearly. There is an urgent need to lift the land-based economy out of the morass it finds itself and make the region more self-reliant than it used to be.

Ladakh currently faces a multitude of challenges related to agriculture: increasing affluent population, rural-urban migration, shrinking agricultural land, shrinking agro-biodiversity, increased dependency on imported food, scarcity during winter, seasonal availability of fruits and vegetables, hidden hunger, post-harvest losses, poor market access, threatened traditional farming cultures and climate change. The challenges are multidimensional and require coordinated and targeted initiatives that can contribute to achieving sustainable agriculture development.

11.1 Growing affluent population

The majority of people in Ladakh live in rural areas. However, the region is experiencing rapid population growth and declining per capita farm sizes. The population of Ladakh has increased from 1,05,292 in 1971 to 2,74,289 in 2011. The urban population of Ladakh saw a rise from 7.5 percent in 1971 to 22.6 percent in 2011. Besides, the lifestyles in villages are also changing fundamentally, i.e., rural roots to urban civilization. In addition to the army, Ladakh's floating population is increasing rapidly. The number of tourists visiting Ladakh increased significantly between 1991 and 2022. Therefore, the influence of population growth on agriculture has a significant impact on the ability of small and marginal farmers to feed themselves and their families. Meeting the increasing requirements of farm produce for the local populace and the floating population in this fragile mountain area is a formidable challenge.

11.2 Rural-urban migration

Rural-urban migration is a global phenomenon. It takes painful dimensions in Ladakh, leaving older people in the villages without helping hands. Small agricultural lands are often abandoned after the migration. The migration would dilute Ladakh's unique culture, which is largely embedded in the rural lifestyle. The rural population in Ladakh was 92.5 percent in 1971, which has come down to 77.4 percent in 2011. The trend cannot be allowed to continue the way it is progressing. Subsistence farming does not offer the conditions and opportunities that young people seek. Agriculture has to become economically viable as well as socially attractive. The shrinking and fragmented agricultural land holdings are one of the reasons for abandoning farming and migrating to urban areas. Other reasons for migration include education, improved social services, and job opportunities. The increasing trend of agricultural land desertion has become a serious threat to farming systems.

11.3 Shrinking agricultural land

Even though there is an increase in agricultural land at the aggregate level, the availability of land for individual farming households has declined in a significant manner. The major decline in land holding is attributed to land fragmentation due to increased population and the breaking up of joint family systems into nuclear families. Ladakh has witnessed a 2.6 fold increase in population between 1971 and 2011. Consequently, the small and marginal holdings of less than 2 hectares of agricultural land account for 79.8 percent of the Leh district and 94.7 percent of the total household in the Kargil district. As a result, small land holdings are becoming non-viable for farming and livelihood. Farmers do not have enough agricultural land to grow substantial crops to meet the food and income needs of the family for a sustainable living.

11.4 Seasonal availability of fruits and vegetables

The climatic condition of the region restricts the crop growing season from May to September in open-field conditions. Vegetables are generally harvested from June to September, while crops such as potatoes, onions, cabbage, carrots, etc., are harvested from late August to September. Fruits are harvested between July to October. The region sees a glut of fruits and vegetables during the period, in contrast to an acute scarcity of fresh produce in winter (Stobdan et al., 2018). During winter, passive solar greenhouse cultivation is the only source of locally produced fresh leafy vegetables. Crops such as potatoes, radishes, carrots, cabbage, onion, and turnip are stored in underground pits and root cellars for their consumption in winter months (Ali et al., 2012).

Radish, cabbage, carrot, potato, onion, leafy vegetables: the only locally produced vegetables for sale in Leh market in winter

11.5 Food dependency

Ladakh is becoming excessively reliant on the outside world for critical needs. It has become a net food importer of cereals, fruits, vegetables, pulses, and cooking oil. The vegetable import dependency in the Leh district is approximately 67 percent, while that of fruit is roughly 85 percent. Self-sufficiency in food is an important issue for the region. Filling the gap between the demand and supply of fresh farm produce is difficult. During winters, a limited quantity of fresh vegetables is brought in by air from Chandigarh or Delhi, paying as much as Rs 110 per kg for air freight. Therefore, fresh vegetables are 2.7-fold costlier in Ladakh compared to the metropolitan city of Delhi (Angmo et al., 2019). Meeting the increasing demand for fresh farm produce at a reasonable price in Ladakh is a formidable challenge.

11.6 Hidden Hunger

Survey results from 200 households, conducted in the year 2008, on assessment of the magnitude of malnutrition and related health problems have found that only 19.4 percent of the people in Ladakh are well nourished. The study also found that 35 percent of people suffer from malnutrition-related diseases. The most prevalent diseases are anemia (8.7), followed by night blindness (7.3), scurvy (6.7), beriberi (6), pellagra (3.7), and rickets (3 percent of the surveyed subjects). A high prevalence of deficiency diseases is attributed to the low consumption of fruits, vegetables, milk, and milk products (Dar and Rather, 2014). A significantly high percentage (36.6 percent) of lactating women in Ladakh are malnourished, as against 19.3 percent of women in Jammu and 10 percent in Kashmir. Similarly, 45.5 percent of Ladakhi women had lower caloric intakes as against 41.3 percent of women in Jammu and 12.7 percent of Kashmiri women. A significant number of the women (36.6 percent) showed clinical signs of nutritional deficiency in Ladakh (Khan and Khan, 2012). Therefore, there is a need to grow more nutritious food locally to make it accessible and affordable.

11.7 Rising cancer cases

The region has a rising trend of cancers and cancer-related deaths. A 10-year (2009–2019) hospital-based study in the Kargil district has found a crude cancer rate of 31.5 cases per lakh population per year. Cancer incidence is higher in males (69.4 percent) than in females (30.6 percent). The pattern of cancers in the region is different from the rest of India, which could be due to the topography, unique food habits, peculiar culture, and lifestyles. Cancer of the stomach is the most common cancer found in 42.1 percent of patients, followed by lung (9.7 percent), liver (9.2 percent), esophagus (5.0 percent), gallbladder (4.5 percent), rectum (2.9 percent), pancreas (2.9 percent), breast (2.5 percent), ovary (2.3 percent) and urinary bladder (2.0 percent). Most of the cancer patients are in the age group of 60–74 (47.3 percent), followed by 45–59 (26.1 percent), 75–89 (14.2 percent), and 30–44 years (5.4 percent) (Hussain et al., 2019). Several studies show an association between a higher intake of certain food and the risk of cancer. A significant association has been found between the consumption of fruits and cancer incidence in Ladakh. Cancer is highest (58.8 percent) in patients taking less than one fruit serving per week, 24.6 percent in those taking one to four fruit servings per week, and only 16.7 percent in those taking more than four per week (Hussain et al., 2019). Therefore, identifying the causes of the increasing number of cancer cases and growing food that could be associated with a lower risk of the disease is a major challenge.

11.8 Poor market access

Inadequate marketing skills and lack of wider market access have been among the biggest problems faced by the farmers in Ladakh, which has a direct bearing on the prosperity of the farmers. Most fruits and vegetable crops attain the harvesting stage in August and September. The perishability of fresh fruits and vegetables makes farmers fearful of possible losses, which may occur during marketing. This prevents them from venturing into horticulture crops. Storage inadequacies result in limited crop production.

11.9 Shrinking agro-biodiversity

The traditional agricultural systems are rich in agricultural biodiversity. Traditionally, people in Ladakh depend entirely on wild harvest collection for their vegetable requirements. Widespread trends towards change in food habits and increased shifts towards cash crops resulted in shrinking agro-biodiversity. Farmers with a market-oriented farming system grow fewer crops and devote less area to mixed crops than traditional farming. Many conventional farming lands are being let permanently fallow and unplanted.

11.10 Limited availability of organic manures

Ladakh is a cold desert with just 141 km^2 of the area under forest cover; the availability of organic matter for farming is a key challenge. The number of livestock populations is also on a decreasing trend.

11.11 Inaccessibility during winter

Due to heavy snowfall, the Ladakh region remains cut off for over five months a year. The region, therefore, suffered from inaccessibility during winter months, which limits the period for the supply of critical inputs.

11.12 Post-harvest losses

Approximately 40-60 percent of the apricot produced in the region is wasted due to a lack of cold-chain infrastructure for storing fresh fruit, hygienic processing, and developing value-added products (Stobdan et al., 2020). The region sees a glut of fruits and vegetables from August to September, which results in post-harvest losses. There is a need for an integrated value chain and market-oriented sustainable system.

11.13 Risk management and resilience

Farmers are exposed to risks from many quarters. Seasonal variability is one such instance in which certain years may not be favorable for fetching a good harvest. The weather affects the yield and quality of the harvest.

Insect pests and diseases can devastate crops. In general, the incidence of insect pests and diseases in the cold arid region is low. However, codling moths, leaf-curling aphids, and gummosis are major problems for fruit trees. *Euproctis similis* is not a regular insect pest of the fruit crop in Ladakh. However, it emerged as a main pest of apricot during 2013–2016 in the region's major apricot growing belts, inflicting a substantial economic loss on the growers (Stobdan et al., 2019). Loose smut is a major disease of cereals in the region, while in the vegetable incidence of the cutworm, onion maggot, aphid, and cabbage butterfly are significant insect pests.

11.14 Threatened traditional farming cultures

Agricultural land has a major influence on social and cultural relevance in the region besides economic and environmental factors. The farming system in the region has evolved from nutritional needs and ecological adaptation. Agriculture and horticulture have shaped and will continue to define Ladakh's unique natural landscape. Even though the trend of cash crops and monoculture began decades ago in and around Leh and Kargil town, it is yet to extend to remote villages of Ladakh. The traditional close ties between agriculture and cultural development can still be seen and felt in Ladakh. However, the traditional farming culture is slowly declining and may face a grave threat of extinction if appropriate actions are not taken now.

11.15 Climate change

Climate change is a global phenomenon, and Ladakh is no exception. The impact of increasing temperatures is now being felt in the region. Studies suggested that the maximum temperature for peak summer months showed a rising trend of nearly 0.5°C between 1973 and 2008, and the minimum temperature in Leh rose by almost 1°C for all the winter months (Angmo and Heiniger, 2009).

We conducted a study in 2016 to gauge people's perception of climate change in Ladakh. We examined how farmers (above 60 years of age) perceived climate change. A structured questionnaire was prepared, and 557 farmers were interviewed. In a startling finding, all respondents stated that they had witnessed climate change in Ladakh, while 74 percent felt that the extent of climate change was extreme. Decrease in snowfall, receding glaciers, change in rainfall pattern, and increase in temperature are perceived as main indicators of climate change in the region. Over 60 percent of the respondents felt climate change severely affected their lives. An increasing number of motor vehicles and urbanization are perceived as the main cause of climate change in Ladakh.

REFERENCES

Ali Z, Yadav A, Stobdan T, Singh SB (2012). Traditional methods for storage of vegetables in cold arid region of Ladakh, India. *Indian J Tradit Knowl.*, 11: 351–353

Angmo P, Dolma T, Namgail D, Tamchos T, Norbu T, Chaurasia OP, Stobdan T (2019). Passive solar greenhouse for round the year vegetable cultivation in trans-Himalayan Ladakh region, India. *Def Life Sci J.*, 4: 103–116

Angmo T, Heiniger LP (2009). Impacts of climate change on local livelihoods in the cold deserts of the Western Indian Himalayan region of Ladakh and Lahaul & Spiti

Dar RA, Rather GM (2014). Assessment of magnitude of malnutrition and related health problems in cold desert Ladakh-India. *Eur Acad Res.*, 2(4): 4895–4919

Hussain S, Ali M, Jeelani R, Abass M (2019). Cancer burden in high altitude Kargil Ladakh: Ten year single centre descriptive study. *Int J Cancer Treat.*, 2(2): 4–10

Khan YM, Khan A (2012). A study on factors influencing the nutritional status of lactating women in Jammu, Kashmir and Ladakh regions. *Int J Adv Res Technol.*, 1 (4): 65–74

Stobdan T, Angmo S, Angchok D, Paljor E, Dawa T, Tsetan T, Chaurasia OP (2018). Vegetable production scenario in trans-Himalayan Leh Ladakh region, India. *Def Life Sci J.*, 3: 85–92

Stobdan T, Chaurasia OP, Wani M, Phunchok T, Zaffar M (2020). Improvement of apricot farming in Ladakh, India, Union Territory of Ladakh, India 50p

Stobdan T, Deen M, Gupta V and Raghuvanshi MS (2019). Integrated management of yellow tail moth (*Euproctis similis*) on apricot in Leh Ladakh India. LAHDC Leh, Ladakh. 50 p.

January 2012: People waiting in long queues to buy fresh vegetables

Chapter-12

Priority Areas to Support Sustainable Agriculture in Ladakh

The priority areas that need support for sustainable agriculture in Ladakh include: harnessing niche potentials, Future Smart Food, diversification towards high-value crops, organic seed production, product differentiation and branding, enhancing agricultural productivity, harnessing water resources, increased agricultural water productivity, increase in crop intensity, agricultural land expansion, protected cultivation, cooperative farming, contract farming, agri-tourism, integrated value chain, marketing, and skilled workforce.

12.1 Harnessing niche potentials

High-mountain products are mainly categorized as specialty or niche products. They are usually produced on a small scale, considering the limited resources available and the size of the rural communities. The high value of these products compensates for the small volumes commercialized. Promoting high-value niche products and services can intensify the total production system. At the same time, the risk of degradation of natural resources or food insecurity can be reduced (FAO, 2019).

The potential for Ladakh agriculture lies in high-value specialty or niche products with high market value. The region is known for its high-quality apricots, apples, and off-season vegetables that can generate significant income for local farmers. The scope also exists for high-quality seed production. However, for various reasons, farmers have yet to be able to harness the local niches provided by the unique agro-climatic conditions.

12.2 Harnessing the potential of Future Smart Food

Future Smart Food (FSF) are neglected and underutilized species that are nutrition-dense, climate-resilient, economically viable, and locally available or adaptable (FAO, 2019). FSF is the key to agricultural diversification in Ladakh. Lakhs of tourists visiting Ladakh every year are the potential targets for popularizing FSF, thus creating extra income and investment for local people. Besides, army and paramilitary forces deployed in Ladakh will also immensely benefit from the FSFs.

12.3 Diversification to high-value crops

The climatic condition of Ladakh offers scope to grow varieties of high-value crops. Crops such as melons, which are believed to be impossible to grow under the cold climatic conditions of Ladakh, have been successfully grown by the farmers. The yield obtained is significantly higher under the organic system than the national average (Angmo et al., 2019). The organic cultivation and the region's unique climatic conditions result in exceptionally high-quality, high-value crops. Encouraging farmers to diversify to high-value crops offers great scope for generating significantly higher employment and helping in increasing farmers' income. Labor intensive nature of growing cash crops can also take care of the problem of unemployment in the region to a great extent.

12.4 Organic seed production

The organic food industry is flourishing. As a result, an increase in farm production of organic food has occurred in many countries. As per regulation, organic farmers must use organic seed material if such is available; as a result, there is a frantic search for certified organic seed by growers. Only a few seed companies have organic seed production facilities, so organic seed is in short supply. With the current trend, the demand for organic cereal, vegetables, and forage seed is expected to increase manifold in the coming years, which requires a massive increase in organic seed production. Ladakh region is an ideal place for seed production due to its low humidity, bright sunshine, and low incidence of disease and insect pest infestation (Stobdan et al., 2018). Ideal climatic conditions for quality seed production and the recent initiative of the government to certify Ladakh as fully organic provide immense scope for harnessing these niches that can hugely benefit the local population. There is a need to invest in organic seed production in the region.

Organic carrot seed production in Ladakh

12.5 Product differentiation and branding

Promoting local products as special and unique is one of the most effective methods of bringing extra value to Ladakh agricultural produce. Branding also helps with consumer recognition of products, ensures higher prices for the same product, and directly accelerates value addition. The pristine environment and rich bioresource will enable products to be sold under 'Brand Ladakh'. The unique plant varieties must carry Brand Ladakh with higher market value than general varieties. For instance, the *Raktsey Karpo* variety became Ladakh's first product to get geographical indications (GI) tag. The *Halman* variety of apricots and the *Karkechu* apple of Ladakh are under consideration for GI tag for their unique properties. The Leh Berry Brand name is a good example of a success story of Seabuckthorn (FAO, 2019). Establishing trusted brands and changing consumer perception towards FSFs is vital at all stages of the process of bringing FSF products to markets. There is a need to promote Ladakh's special and unique products in national and international agricultural exhibitions, which will undoubtedly increase the visibility and promotion of the brand and sales of the produce. There is a need to provide regulatory support such as standards, certification, branding, labeling, and implementing sustainable management regimes for Ladakh specialty products.

12.6 Investment and entrepreneurship

A lasting agriculture transformation is one that is ultimately supported by real market forces. Introducing innovation requires a market stimulus to induce potential entrepreneurs and investors to take on a defined set of initiatives. Identifying incentives and building infrastructure will kick-start and accelerate investment in the agriculture and horticulture sector. The enabling environment may be created for market-driven economic growth and investment.

Emphasis may be accorded to implementable policies that attract the resources and private sector players into investing in commercial agriculture, ensuring the inclusiveness of small landholders, women, and the youth. In addition, integrating the different sectors of the region's economy with agricultural production systems, the flow of credit to rural areas, and increased investment in agricultural enterprises will help make all farming systems economically viable and ecologically sustainable.

12.7 Integrated value chain-based and market-oriented sustainable system

The integrated value-chain development approach is a market-driven systems approach that focuses on linking households and/or communities to growing markets so that they can earn income to purchase additional food while reducing the risks of relying solely on their production. In countries such as China and Nepal, integrated value-chain development has brought many positive effects, which include responsible production, environmental protection, and sustainable development. It creates shared value by strengthening producers and their groups and organizations to collectively acquire agricultural inputs at lower prices, helping them add value and commercialize their products (FAO, 2019).

A huge opportunity lies in the processing and value-addition of agricultural and horticultural produce. Processing offers an instant gateway for entrepreneurs into the sector, allowing them to connect to farming communities, drive agricultural development, and directly impact the livelihood of farmers. Approximately 40–60 percent of the apricot produced in the region is wasted (Stobdan et al., 2020). Less than 5 percent of the Seabuckthorn in the region is harvested primarily because of a lack of an integrated value chain and market-oriented sustainable system. Over 90 percent of Seabuckthorn harvested in Ladakh is sold after primary processing without any value addition (Stobdan and Phunchok, 2017).

Strengthening the local entrepreneurs to set up small and medium food processing units in different parts of

Ladakh will strengthen the integrated value-chain development in the region. The region must create an enabling environment and promote an integrated value chain-based and market-oriented sustainable system. Recently, DRDO and UT Ladakh jointly established an apricot processing plant in Leh, which is run under GOCO (Government-owned company-operated) model. Similar models need to be established in major production areas for apricot, apple, Seabuckthorn, milk, and pashmina.

12.8 Marketing

Lack of organized agricultural marketing has been among the biggest problems faced by the farmers in Ladakh, which has a direct bearing on the prosperity of the farmers. Most fruits and vegetable crops attain the harvesting stage in August and September. The perishability of fresh fruits and vegetables makes farmers fearful of possible losses, which may occur during marketing. This prevents them from venturing into farming. Storage inadequacies result in limited crop production. The potential of the local market is often underestimated. The local market is flooded with imported farm produce, most of which can be grown or processed in Ladakh. Imported dried apricot is one such product that can easily be replaced in the local market. Similarly, the local market is flooded with *Rajma* (kidney beans) despite Ladakh being known for producing high-quality *Rajma*. Organizing the collection, processing, value-addition, and marketing of such farm produce is necessary. The local market offers substantial opportunities for commercial farming.

12.9 Agri-tourism

Tourism has emerged as one of the dominant components of the current Ladakhi economy. However, the benefits of the tourism industry are still concentrated mainly in and around Leh. Little of this income reaches remote communities. Agricultural tourism is a worldwide trend that offers city dwellers a chance to escape urban concrete and re-discover their rural roots. Farmers in Ladakh can effectively participate in several ways in agriculture tourism. In the last two decades, homestay ecotourism, hosting tourists as guests in their homes, has been gaining popularity in Ladakh. Homestays were proposed in 2000 by the Snow Leopard Conservation Trust of India as a means of providing an alternative income for farmers whose livestock were threatened by snow leopards, with the hope that farmers would see these highly endangered predatory cats as a source of income rather than a pest needing to be eradicated. The idea was first introduced to villagers in the community of Rumbak (Peaty, 2009) and later spread to many other villages.

12.10 Cooperative farming

To overcome the limitations of small farm size and limited production, cooperative approaches to business conduct will help achieve sustainable development. Small household economies can be converted into larger units at the village level, managed by trained experts from within the village or hired workforce. Small and marginal farmers may not always have the means of transportation for delivering their products to the markets, which puts them into an unfavorable negotiating position with respect to intermediaries and wholesalers. A cooperative can act as an integrator, collecting the output from members, sometimes undertaking to manufacture, and delivering it in large aggregated quantities downstream through the marketing channels. A pool of agricultural machinery also needs to be created where a small family farm does not justify owning expensive farm machinery; instead, such farmers may form a machinery pool to acquire the necessary farm equipment for the use of its members. Such programs have helped China revitalize the rural mountain economy dependent on small and marginal farms by helping convert these into profitable farming enterprises (IIM & FAO, 2019).

12.11 Contract farming

Contract farming is another means to secure demand for mountain products. Through contract farming, farmers can increase on-farm income and diversify livelihood strategies. In a local context, where farmers lack alternatives for income generation from agricultural production, the benefits of contract farming with a guaranteed purchase without transportation costs improve their market access and help generate on-farm income (FAO, 2019).

A variety of off-season vegetables are being grown in the region that can be marketed to supermarkets, specialized wholesalers, processors, and agro-exporters through contract farming (Stobdan et al., 2018). PepsiCo India Holdings Private Limited was the first (inter-)national company to introduce contract farming in the region in 2007. In 2008 farmers in Ladakh received Rs

70 lakh (approx) from the sale of chip-grade potatoes to PepsiCo. Although farmers' expectations on yield and returns have been fulfilled, the contract stopped after two years primarily due to the high cost of production. In 2009 AVT Natural Product Ltd. Cochin initiated pilot-scale contract farming for marigold seed production in Ladakh. Dabur India has also entered into contract farming of Pushkarmool (*Inula racemosa*) cultivation. DRDO-DIHAR facilitated the above three pilot-scale contract farming. There is immense scope for contract farming in the region to boost the region's economy. There is a need to frame support and policies to institutionalize contract farming in the region.

12.12 Protected cultivation

The temperature in Ladakh often drops down to -30°C in winter. Harsh winters reduce the cropping season to four to five months a year. The region remains cut off for over six months in a year due to heavy snowfall. Emphasis needs to be given to promoting protected cultivation at a massive scale to intensify the production of high-value crops and increase water use and production efficiencies per unit area. Adopting protected cultivation at a massive scale will meet the regional requirements of fresh fruits and vegetables, and has an immense potential to sell the product outside the region. There is a need to invest in bringing large areas under protected cultivation.

12.13 Enhancing agricultural productivity

The productivity of most of the crops in the region is low, and there is considerable scope to increase it. Even within the region, there is a large variation in yield due to variations in climate, soil, elevation, availability of inputs, and access to irrigation. Lower yield in the region may be due to low adoption of improved technology. Enhancing access to affordable technological advancement is the most potent instrument to raise agricultural productivity in the region. Ladakh agriculture has a unique set of challenges. Considering the limited area of cultivation in the region, vertical expansion in terms of productivity gains through agroecological intervention and sustainable practices will help to enhance food and nutrition security. There is a need to acquire the improved technology packages and make them available to the farmers and producers to bridge the gap between current low productivity to its full potential. This is feasible through the strengthening of specialized well-targeted research and extension.

12.14 Increase in crop intensity

For making agriculture profitable on small and marginal farms, intensive agriculture holds the key for them. Single-cropping is dominant in Ladakh due to the short agricultural season. However, double-cropping is practiced in limited areas that fall below an altitude of approximately 3000 m asl due to warmer climatic conditions. Two crops are taken in a year on the same land in such areas. Traditionally buckwheat is grown as a second crop after harvesting barley crops. With the availability of new technologies and crop varieties, raising short-duration crops after the main crop has become possible. Taking the second crop, such as turnips and leafy vegetables, on the same piece of land is a significant source to address the region's agricultural land constraint and raise income per unit of land. In some vegetable growing areas, the cropping intensity has increased to 300 percent under greenhouse conditions. There is a need to explore technologies and crops for increasing crop intensity in the region.

12.15 Increased agricultural water productivity

The Government is committed to accord high priority to water conservation and its implementation. To this effect, *Pradhan Mantri Krishi Sinchayee Yojana* (PMKSY) has been formulated with the vision of extending the coverage of irrigation (*Har khet ko pani*), improving water use efficiency (More crop per drop) in a focused manner with an end-to-end solution on source creation, distribution, management, field application, and extension activities. This scheme needs to be accorded top priority in Ladakh in view of its immense importance in managing scarce water resources to increase agricultural production.

The future direction and success of Ladakh farming will depend on water availability, particularly the ability to increase water productivity. There is so much to learn from Israel's irrigation practices. The geography and climate of both Ladakh and Israel are not naturally conducive to agriculture. Both regions are largely desert, and the lack of water resources does not favor farming. The Israeli agro-technology needs to be tailored to the local conditions.

12.16 Harnessing water resources

Although mountains provide freshwater resources to 50 percent of the world's population, people living in mountain regions face water scarcity during the dry months of the year. This contributes to both food insecurity and poverty (FAO, 2019). Glacier-fed and snow-fed gravity-controlled irrigation systems are predominant in Ladakh. Adequate meltwater supply from glaciers or permanent snowfields in the upper catchments or, where local topographical conditions permit, direct abstraction from the mainstream is a fundamental prerequisite for crop production in Ladakh. Glaciers provide most of the runoff in summer, with a maximum regularly reached in the afternoon. In such water surplus situations, the evening and nighttime runoff is usually stored in reservoirs (*Zing*) and diverted to the fields in the morning when the natural runoff is scarce (Nüsser et al., 2012). Severe water scarcity occurs during the sowing period in April and May due to snow-deficient winters or delayed snowmelt. A sufficient and reliable meltwater supply from high-altitude glaciers becomes available from June to October.

Integrated watershed management: In 1996, the Watershed Development Programme was introduced through the Desert Development Programme in selected villages in Ladakh. The lessons learned from the previous program provide valuable new understandings that can help in future projects.

Lift irrigation: In Ladakh, the river water still needs to be utilized for irrigation purposes. Even villages that are located just 100 m away from the river do not get water from the river simply because it is located too high to permit irrigation by gravity flow from the source. Lift irrigation is being increasingly practiced in India, and the technology can easily be adopted in the region. River water may be lifted by installing solar or electrical pumps in the main delivery chamber, which is situated at the topmost point in the command area. The water can then be distributed to the field of the beneficiary farmers using a suitable and proper distribution system. The requirements for a viable lift irrigation scheme are a constant water source for the whole irrigation season at the site and the feasibility of lifting water to the desired location.

The lift irrigation system can be owned individually or by a group of farmers in a cooperative mode. The government-run lift irrigation project in Phey village in Ladakh has been running successfully for decades. The village is located next to the river Indus but has to deal with recurrent water scarcity, especially during the sowing season, due to its dependence on meltwater from glaciers. The lift irrigation project made water available during the early cropping phase, making growing melons and vegetables possible in the village. Seeing the positive impact of the government-run water lift project, the villagers launched their community- and individual-owned water lifting system. Similar projects must be replicated in villages that have to deal with recurring water scarcity. Besides, the vast barren land on both sides of the major rivers needs to be put under cultivation by lifting the river water. Solar pumps need to be used extensively to cut down recurring expenditures.

Artificial glaciers: Artificial glaciers, located at much lower altitudes than the naturally occurring glaciers above 5200 m asl, bridge the critical water availability gap during the sowing season (April-May) by providing ice reservoirs that melt earlier in the agricultural season. Despite the popularity of the term 'artificial glacier', the term 'ice reservoir' conveys their character and function more accurately because the ice bodies only exist for a few months, unlike glaciers in the strict sense, which are defined as perennial, moving ice bodies with distinct accumulation and ablation zones. Such ice reservoirs utilize the hydrological process of icing under local conditions of frequent freeze-thaw cycles to capture water for seasonal storage. They are not water storage structures that freeze from the top down; rather, they are produced through the sequential freezing of thin layers of water, creating superimposed sheets of ice. These ice reservoirs are maintained as communal infrastructure reliant on local institutions and external technological interventions (Nüsser et al., 2019). Such projects need to be promoted to harness the water resources in the remote villages in Ladakh.

12.17 Farm mechanization

Agricultural mechanization plays a vital role in optimizing the use of land, water, human resources, and other inputs to maximize the productivity of the available cultivable area and make agriculture a more profitable and attractive profession for rural youth. Since most farmers have small land holdings, the schemes for establishing Custom Hiring Centres and Farm Machinery Banks must be promoted in the region.

12.18 Agricultural land expansion

Only 0.4 percent of Ladakh's geographical area is used in crop production. Agricultural land expansion is a significant factor in increasing crop production and livelihood improvement in Ladakh. Large tracts of vacant arable land can be brought into cultivation with minimal environmental impact. The primary constraint in area expansion under crop production is the assured water source for irrigation. There is immense scope to expand the area from the existing 23,612 hectares. With technological advancement, lifting water from the river for irrigation is now feasible. The glacial water that drains into the rivers can be conserved and judiciously used to bring more area under cultivation. Various technologies are available for improved operation, management, and efficient irrigation water use. There is a need to initiate a myriad of regionally targeted initiatives to expand agricultural land in Ladakh.

12.19 Human capital

Attracting the required skills and talent is crucial in realizing the goals of thriving agriculture and horticulture sectors. The agricultural and horticultural sector must be able to plan and secure the appropriately skilled people it needs across the permanent and seasonal workforce. Currently, a significant proportion of the agricultural workforce comes from outside Ladakh.

Agriculture and horticulture are increasingly becoming high-tech, capital-intensive industries. Presently knowledge deficit exists. Farmer's exposure to sophisticated technologies and skill development programs will encourage them to adopt the latest tools and technologies and develop the right agronomic and business management skills appropriate for the region. The current trends indicate that agriculture in Ladakh needs to adopt smart agriculture technologies that will enable significantly greater productivity while dealing with an extensive scarcity of resources, including human resources, to achieve sustainable agricultural production. Agriculture needs to emerge as an attractive career for the youths of Ladakh. There is a need for investment in human resource development to make agriculture an exciting career path for the youths.

12.20 Supporting farming in less favored areas

Farming activity in remote and less favorable areas, such as extremely high altitude conditions, needs support through special schemes. These areas have fewer opportunities to improve their productivity and fewer opportunities to market their products and diversify. However, these areas are unique in their terms, and support is required to encourage preserving their traditional farming system, encourage biodiversity, conservation of water sources, and protection of wildlife.

12.21 Strengthening traditional farming system

Undertaking agricultural reforms in selected areas to strengthen and enhance the traditional farming system in the region will go a long way in preserving the age-old practices. Price support and special incentives may be given to farmers for practicing traditional farming systems. This will ensure that the age-old traditional agricultural system is practiced, which in turn helps in environmental conservation and promotion of agri-tourism in the region.

12.22 Technology back support

The climatic condition of Ladakh is unique, and accordingly, the innovations, agro-practices, plant protection measures, and technological inputs must be tailored to suit regional requirements. Simply relying on imported knowledge and materials is not the solution for a high-altitude sustainable farming system. Expertise must be developed in-house to consider the potential value of the region's niches and resource base. The Ladakh-based research institutes–Defence Institute of High Altitude Research (DIHAR), Regional Station of Sher-e-Kashmir of Agriculture Science & Technology SKUAST (K), Regional Station of Central Arid Zone Research Institute (CAZRI), Regional Centre of G. B. Pant National Institute of Himalayan Environment (GBPNIHE) – can play vital roles in developing expertise to offer technology backed support to the farming community.

REFERENCES

Angmo S, Dolkar D, Dolkar P, Kumar B, Stobdan T (2019). Growing watermelon in high altitude trans-Himalayan Ladakh. *Natl Acad Sci Lett.*, 42: 379–382

FAO (2019). *Mountain agriculture: Opportunities for harnessing Zero Hunger in Asia*. Bangkok.

IIM & FAO (2019) *State of the Himalayan farmers and farming.* India.

Nüsser M, Dame J, Kraus B, Baghel R, Schmidt S (2019). Socio-hydrology of 'artificial glaciers' in Ladakh, India: assessing adaptive strategies in a changing cryosphere. *Reg Environ Change*, 19:1327–1337

Nüsser M, Schmidt S, Dame J (2012). Irrigation and Development in the Upper Indus Basin. *Mt Res Dev.*, 32: 51–61

Peaty D (2009). Community-Based Tourism in the Indian Himalaya: Homestays and Lodges. *J Ritsumeikan Soc Sci Humanit.*, 2: 25–44.

Stobdan T, Angmo S, Angchok D, Paljor E, Dawa T, Tsetan T, Chaurasia OP (2018). Vegetable production scenario in trans-Himalayan Leh Ladakh region, India. *Def Life Sci J.*, 3: 85–92

Stobdan T, Chaurasia OP, Wani M, Phunchok T, Zaffar M (2020). *Improvement of apricot farming in Ladakh, India*, Union Territory of Ladakh, Administration of UT Ladakh, India, 50 p

Stobdan T, Phunchok T (2017) Value chain analysis of Seabuckthorn (*Hippophae rhamnoides* L.) in Leh Ladakh, Ministry of Agriculture and Farmers Welfare, Govt of India

Chapter-13

Priority Crops for Sustainable Agriculture in Ladakh

A variety of crops are being grown in Ladakh. As many as 101 vegetables can be grown in the field in a single season (Stobdan et al., 2018). However, most of the crops are produced on a small scale, considering the limited resources available and the size of the rural communities. Besides growing crops for own consumption, the region has the potential to grow high-value specialty or niche products that have a high market value. Ladakh must prioritize growing only those crops commercially for which the region has a competitive edge. The GI tag for *Raktsey Karpo* variety of apricot and the maiden export of apricots in 2021 is an eye-opener for the people of Ladakh. The region is known for its high-quality apricots, apples, and off-season vegetables that can generate significant income for local farmers. The scope also exists for high-quality seed production. Ladakh also has a strong niche potential for organic agriculture and livestock production. However, for various reasons, farmers have yet to be able to harness the local niches provided by the unique agro-climatic conditions.

13.1 Apricot

Ladakh is the biggest apricot producer in India, with a total production of 15,864 tonnes. Apricots of Ladakh are known for their quality. Historically, dried apricot was one of the four natural products of Ladakh that were traded with neighboring countries (Cunningham, 1854). *Phating*, Ladakh's premium quality dried apricots, is known as *Nyari Khambu* in Tibet. We found that *Raktsey Karpo*, apricots with white seed stones, is unique to Ladakh. It is the most preferred cultivar for fresh consumption (Naryal et al., 2019). Apricots of Ladakh are harvested between mid-July and early September and have a distinct competitive advantage as it does not coincide with the main apricot season in the market (Naryal et al., 2020). In recent years the apricots of Ladakh have been receiving attention from the government and the industry. The maiden export of Ladakh apricot began in 2021 to Dubai. Approximately 20 tonnes of fresh apricots were sold in Mumbai, Hyderabad, Ahmedabad, and Delhi in the same year. During the 2022 season, 35 tonnes of fresh apricots were sent to the domestic market outside Ladakh and exported to Singapore, Mauritius, and Vietnam. Ladakh could emerge on the world map for quality apricot production. The premium quality *Halman* and *Raktsey Karpo* cultivars have immense potential to change the region's economy.

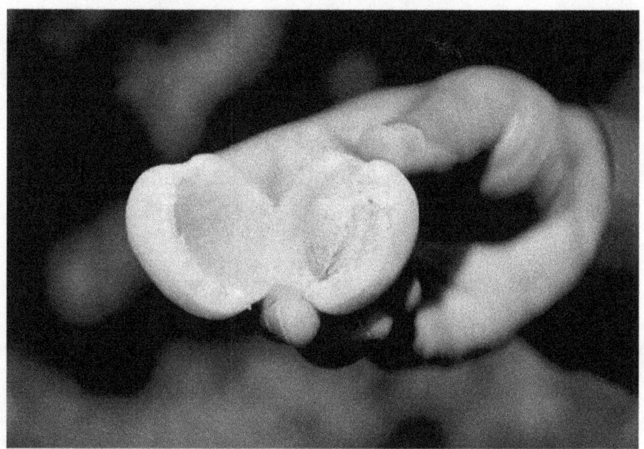

Raktsey Karpo: apricot with white seed stone, unique to Ladakh

13.2 Apple

Reports suggested climate change has affected apple production in major apple-growing regions in India. Besides the yield loss, the quality of fruit produced in these areas has also degraded (Sharma et al., 2017). Given the changing scenario, the high-altitude Ladakh region has the potential to emerge as a favorable place for quality organic apple production (Dolker et al., 2023). The region experiences long day hours with high light intensity, warm days with cool nights, and low relative humidity from May to October, thus making apple production favorable in the region. The red apple cultivars formed intense red skin fruits naturally. The apples produced in Ladakh are late maturing with high total soluble solids (TSS), intense colored skin with high phenolic contents, and has a comparative advantage as that of major apple-growing regions of India. The climatic conditions and topography of the region are favorable for growing organic apples with a low incidence of insect pests and diseases. *Karkechu* apple has been identified as unique to Ladakh for its strong fruit aroma, and we filed an application for GI tagging. Based on our extensive research at DRDO-DIHAR, we recommend *Karkechu*, *Thra*, Gale Gala, Golden Spur, and Royal Delicious for the region.

13.3 Seabuckthorn

Seabuckthorn (*Hippophae rhamnoides* L.) is an ecologically and economically important thorny shrub that is widely distributed in the Ladakh region. Ladakh remains the major site (13,000 ha) for natural Seabuckthorn resources, with over 70 percent of the total area in the country. Once considered a thorny menace, Seabuckthorn is now being seen as a means for the region's sustainable development. Farmers and processors in Ladakh are getting incentives from the Horticulture Department for growing and processing Seabuckthorn. Centre of Excellence on Seabuckthorn has also been sanctioned under the MIDH scheme. The Seabuckthorn harvest has doubled between the years 2018 and 2021. In 2021, 985 tonnes of Seabuckthorn were harvested, and locals got Rs 45–55 per kilogram of wild fruit. The pulp is sold at Rs 175 per liter, and most of the harvest is currently exported.

13.4 Cherry

Cherry is one of the pricier temperate fruits but not a traditional fruit crop of Ladakh. My colleagues and I have conducted studies and found that cherry has immense potential as a new cash crop for the Ladakh region. Appearance is the primary criterion in purchasing decisions, and cherries produced in Ladakh developed intense red color due to bright sunlight and UV radiation. In general, full dark red cherries have higher consumer acceptance than full bright red cherries. Cherries are the first fruit to ripen in Ladakh, usually in July, and can be marketed as off-season fruit.

13.5 Peach

Peach is the most important temperate fruit crop worldwide in terms of production after the apple. It is revered as a delicious and healthy summer fruit in most temperate regions worldwide. It is highly perishable, but depending on market demands and availability, it can be a profitable fruit crop for the growers. Peaches are traditionally grown in Ladakh and known as '*Tra Kushu*'. However, the fruits of the traditional genotypes are small in size and not intensely colored. We at DRDO-DIHAR studied elite exotic cultivars with enhanced aroma and color with appreciable nutritional properties. The cultivar Suncrest performs very well in Ladakh. The fruits are ripe in August-September and thus have a competitive advantage as off-season fruit.

13.6 Watermelon

Watermelon (*Citrullus lanatus*) is a warm-season crop whose global consumption is greater than any other cucurbit. After many years of experimental trial, we recently introduced watermelon as a new cash crop for Ladakh (Angmo et al., 2019). Fruits are harvested in August and September and thus can be marketed as off-season fruit. Watermelon is now being grown at a commercial scale under open-field conditions. Farmers earn Rs 10–12 lakh per hectare, which is approximately ten times more than cereal crops.

13.7 Sun melon

Sun melon (*Cucumis melo* var. *inodorous*) is a specialty melon. It is highly relished because of its attractive fruit with a unique flavor and sweet taste. Sun melons' increasing popularity and market share attracts farmers to grow this fruit type. However, sun melon is a warm-season crop traditionally not grown in the trans-Himalayan region. The local requirement is transported from outside the region. We at DRDO-DIHAR conducted studies with ICAR-Indian Agricultural Research Institute (IARI) and found that sun melon could be successfully grown under an organic farming system. Marketable fruit yield of 25-35 tonnes per hectare is obtained under plastic mulching in open field conditions. Fruits are harvested in August and September and thus can be marketed as off-season fruit.

13.8 Off-season vegetables

Ladakh has advantages over the plains due to its topography and agro-climatic conditions. Off-season vegetables that cannot be produced in the plains during the summer are being grown. Turnip is among the first crops to harvest in the region from May onwards. Most crops reach the harvestable stage from late July onwards and are harvested before October (Stobdan et al., 2018). Various off-season vegetables, such as broccoli, cabbage, cauliflower, peas, etc, are being grown that can be marketed to other parts of the country. Disease and insect infestation are also minimal; therefore, vegetable products are free of chemical pesticides. There is immense scope for harnessing these niches to benefit the local population.

13.9 Cut flowers

Ladakh's climatic condition is ideal for producing off-season cut flowers. The prevailing low temperature, photosynthetic irradiance, and high UV-B are known to cause strong flower color. Common in Ladakh, water deficiency also causes the flower to turn darker. Therefore, the flowers grown in Ladakh develop strong colors, an important determinant that directly influences their commercial value. Commercial cultivation of gladiolus began in the mid-1990s in Saboo village, and about 15 ha of the area was under gladiolus (Dwivedi et al., 2002). Lilium is another important cut flower grown in Ladakh. In August of 2022, the first consignment of lilium, consisting of approximately 800 flower sticks, was shipped for the first time from Ladakh to Delhi under CSIR Floriculture Mission with the support of the Department of Industries and Commerce, UT Ladakh. Growing cut flowers in the open field can increase farm income. However, the market linkage is the deciding factor for the success of the commercial cultivation of lilium.

13.10 Flower bulb production

The flower bulb sector produces and trades bulbous and tuberous plants, rhizomes, and root tubers. The climatic condition of Ladakh is favorable for the production of flower bulbs. The low incidence of insect pests and diseases is an added advantage for the region. However, the sector still needs to be addressed, and the region has immense scope for flower bulb production.

13.11 Buckwheat

Buckwheat (*Fagopyrum esculentum*) is grown traditionally in relatively warmer areas of Ladakh, where double-cropping is possible. It is generally grown as the second crop after harvesting the barley crop. Buckwheat is locally known by various names, viz. *Dyat, Dro, Bro, Fafar,* etc., in different regions of Ladakh. Two variants of buckwheat are grown in Kargil: yellow-colored, small-sized *Brosuk*; and black-colored, larger-sized *Gyamrus*. It is less productive on good soil than other grain crops but particularly adapted to poor, badly tilled land. It is one of the quickest-growing green manure crops, taking only 4–5 weeks from seeding to flowering (Ahmed and Raj, 2012).

13.12 Goji berry

Gogi berry (*Lycium ruthenicum* Murray) is locally known as '*Khizer*' in Ladakh. The fruits of *Lycium* genus are used to treat many health conditions, including early-onset of diabetes, anemia, vision problems, impotency, and lung disorders. Goji berry is commonly found in sandy places of Hunder and Diskit areas of Nubra Valley at an altitude of about 3,100 m above msl. The dark violet-colored fruits contain water-soluble edible dye, which the traditional *Amchis* have used in various herbal formulations as medicine or as a coloring agent (Dhar et al., 2011). The potential of the crop as a major commercial crop is underway.

13.13 Wild rose

Wild rose (*Rosa webbiana* Wall. ex Royle) is locally known as '*Sia*' in Ladakh. It is drought-tolerant and grows even in extremely high-altitude areas in Ladakh. The name of the Siachen Glacier is derived from the wild rose. '*Sia*' in the local dialect refers to rose, and '*Chen*' refers to any object found in abundance. Thus, the name Siachen refers to a land with abundant roses. The petals are used as an important ingredient in preparing herbal tea developed by DRDO-DIHAR. The hips of *R. webbiana* are edible and are rich in Vitamin C. There are no specific examples of roses being cultivated in India for their hips (Tejaswini & Prakash, 2005); therefore, wild roses can be grown on a large scale in Ladakh for hips and petals.

REFERENCES

Ahmad F, Raj A (2012). Buckwheat: a legacy on the verge of extinction in Ladakh. *Current Sci.*, 103: 13

Angmo S, Dolkar D, Dolkar P, Kumar B, Stobdan T (2019). Growing watermelon in high altitude trans-Himalayan Ladakh. *Natl Acad Sci Lett.,* 42: 379–382

Cunningham A (1854). Ladakh: Physical, statistical, and historical with notices of the surrounding countries. Gulshan Publisher, Gow Kadal, Srinagar, India

Dhar P, Tayade A, Ballabh B, Chaurasia OP, Bhatt RP, Srivastava RB (2011). *Lycium ruthenicum* Murray : a less-explored but high-value medicinal plant from the trans-Himalayan cold deserts of Ladakh, India. *Plant Arch.*, 11: 583–586

Dolker T, Dolma T, Kumar Deepak, Sharma NC, Chaurasia OP, Stobdan T (2023). Growing organic exotic apples in trans-Himalayan Ladakh in climate change scenarios. *Proc Natl Acad Sci India Sect B Biol Sci.*, DOI: 10.1007/s40011-023-01489-w

Dwivedi SK, Paljor E, Attrey DP (2002). Gladiolus in Ladakh. Field Research Laboratory, Leh Ladakh

Naryal A, Angmo S, Angmo P, Kant A, Chaurasia OP, Stobdan T (2019). Sensory attributes and consumer appreciation of fresh apricots with white seed coats. *Hortic Environ Biotechnol.*, 60: 603–610

Naryal A, Dolkar D, Bhardwaj AK, Kant A, Chaurasia OP and Stobdan T. 2020. Effect of altitude on the phenology and fruit quality attributes of apricot (*Prunus armeniaca* L.) fruits. *Def Life Sci J.*, 5: 18–24.

Sharma DP, Sharma HR, Sharma N (2017). Evaluation of apple cultivars under sub-temperate mid-hill conditions of Himachal Pradesh. *Indian J Hort.*, 74:162-167

Stobdan T, Angmo S, Angchok D, Paljor E, Dawa T, Tsetan T, Chaurasia OP (2018). Vegetable production scenario in trans-Himalayan Leh Ladakh region, India. *Def Life Sci J.*, 3: 85–92

Tejaswini, Prakash MS (2005). Utilization of wild rose species in India. *Acta Hort.*, 690: 91–95

The Receptiveness of Ladakhi Farmers Towards the Adoption of New Agro-technologies

The diffusion of new technologies is a crucial contributor to economic growth, and differences in technology use account for much of the inequality. Farmers' decision to adopt new agricultural technology in preference to old technologies depends on complex factors such as appropriateness in terms of farmers' needs, socioeconomic status, the psychological makeup of the individual, resource availability, demographic characteristics, and access to institutional support. Due to the single cropping season, perceived risk is a significant factor in the decision-making process for adopting new technologies in Ladakh (Dorjey and Srivastava, 2013). Understanding the relationship between these factors and the new technology adoption process is essential. There is considerable evidence showing that farmers generally have a subjective preference for agro-technologies, which significantly affects adoption decisions. The farmers sometimes reject technologies that work well on research farms. The advice and information provided are sometimes rejected since the farmers cannot process information effectively and/or are not receptive to advise. Adopting new technology is shaped by the advice-giver identity and the receiver's receptiveness. Adopting the 'copy and paste' model of technology intervention, where models adopted for lowlanders are applied to the Ladakh region, is ineffective.

The dissemination of regionally suited technologies to the local farmers via the widespread network facilities of the UT Departments of Agriculture, Horticulture, and various NGOs is effective in the region. These networks are spread at different grassroots levels and directly supply inputs to the local farmers (Angchok and Srivastava, 2013). To effectively disseminate the new technologies, it is crucial to understand the mechanism followed by the UT Agriculture and Horticulture Departments to reach the farmers at the grassroots level through various schemes. Unless a new technology becomes a part of the departmental scheme, its spread to far-flung villages is most unlikely. In recent years, the region has witnessed many success stories of effective dissemination of regionally suited technologies developed by Ladakh-based research institutes. Sufficient time, typically two to three years, is required for the effective diffusion of new technologies among the early adopters. Once selected villages or progressive farmers adopt the technology, it diffuses among most farmers through various schemes of the Agriculture and Horticulture Departments.

14.1 Growing watermelon as a cash crop

Diversification towards high-value crops offers excellent scope for generating significantly higher employment and helping to increase farmers' income. Accordingly, my colleagues and I at DRDO-DIHAR developed agro-techniques for growing watermelon, a warm-season crop, as cash crop in open fields in Ladakh (Angmo et al., 2019).

We conducted the first on-farm demonstration of growing melons on farmers' fields with the Agriculture Department in Phey village in 2016. Similarly, we also conducted trials in collaboration with LEHO, an NGO, on fields of one farmer each in Dha, Lasthang, Achnathang, Domkhar Dho, Domkhar Barma, Takmachik, and Hanuthang villages. The trials were a big success, and farmers could replicate the results obtained at our research farm. When we repeated the trials in 2017, LEDeG, an NGO, joined the field demonstration trials. Upon completing a successful two-year trial in the farmers' fields, DRDO-DIHAR formally transferred the technology to the Agriculture Department, LAHDC Leh, on 06 October 2017. The Agriculture Department, since then, has been promoting growing watermelon in the region under various schemes. Watermelon is now grown commercially under open-field conditions in 55 villages in the Leh district.

14.2 Mulching for doubling vegetable productivity

The productivity of most of the crops in the Ladakh region is low, and there is considerable scope to increase it. Lower yield in the region compared to national averages are due to harsh climatic conditions, poor level, or low adoption of improved technologies. Fortunately, our many years of research show that using black plastic mulch in the region improves farmers' income and rural livelihoods by increasing productivity and profitability. Over two-fold increase in tomato, capsicum, brinjal, and okra crop productivity has been demonstrated. The research and assessment results showed that adopting the technology can enhance farmers' income growth. The farmer-friendly technical know-how was formally transferred by DRDO-DIHAR to the Agriculture Department, LAHDC, on 12 March 2015 for large-scale adoption at the grassroots level. The benefit of the technology was demonstrated on farms to the Agriculture Department in 2015.

We conducted the first demonstration on using black mulch in farmers' fields in Phey village in 2016. Farmers were asked to grow tomato crops with and without mulching. Even though the farmers have not used mulching technology before the trial, most of the farmers found the use of plastic mulch either very easy (86 percent) or easy (14 percent) during the first year of the trial. All the farmers followed the package of practices recommended during the trial. As suggested before the trial, farmers have also compared the performance of tomato crops under mulch with the traditional cultivation methods. Some farmers also tried crops such as capsicum, cucurbits, and brinjal under mulching. These crops were not suggested for the trial, but the positive effect of the use of black plastic mulch on these crops was highlighted during the training program. Therefore, farmers in Ladakh are risk-takers and readily adopt new technologies.

Feedback from the farmers after the first trial suggested that weed suppression was ranked as the main benefit of mulching by 51.8 percent of farmers. Farmers were able to get double the yield as compared to bare soil. An increase in fruit yield was regarded as the main benefit of mulching by 19.1 percent of farmers. Early harvest was ranked as the main benefit of using plastic mulching by 16.6 percent of respondents. The remaining 12.7 percent of farmers regarded water saving as the main benefit of mulching (Angmo, 2019). Doubling productivity in crops such as tomato, capsicum, brinjal, and okra has become a reality in Ladakh. It is also used for growing melons in the region.

The use of black plastic mulch is widespread in the region for increasing yield, water conservation, and early crop harvest. The technology is included in the District Plan of Agriculture Department, LAHDC Leh, from 2016–17 onwards, and it has been adopted in 72 villages in the Leh district.

14.3 Ladakh Greenhouse

Many passive solar greenhouses have been established in the region since the 1980s. However, given the limitations of the traditional greenhouses, we at DRDO-DIHAR developed an improvised passive solar 'Ladakh Greenhouse' technology, wherein growing cauliflower, cabbage, broccoli, tomato, chili, cucumber, and mushroom in sub-zero temperatures was demonstrated, which otherwise is not possible in the traditional passive solar greenhouses. The greenhouse works all year round without a supplementary heating or cooling system.

The first trial of the greenhouse on farmers' fields began in 2020. Initially, five greenhouses were established by DRDO-DIHAR in Thiksey village to demonstrate the working of the new technology on farmers' fields. On successful demonstration of the technology, Ladakh Greenhouse technology was adopted by UT Ladakh for the large-scale establishment on farmers' fields. Over 1500 Ladakh Greenhouses have been established in 120 villages within two years, and more greenhouse construction is underway.

Farmers in Ladakh grow cauliflower, cabbage, broccoli, tomato, chili, and cucumber in peak winter months, which is otherwise is impossible in the traditional passive solar greenhouses. Besides alleviating hidden hunger, a farmer earns Rs 60,000–80,000 per year from a single greenhouse. Due to its uniqueness and impact on the grassroots level, the Ladakh Greenhouse technology received the Prime Minister Award for Excellence in Public Administration (under the Innovation category) on 21 April 2022. The UT administration provides 75 percent financial assistance to farmers to establish Ladakh Greenhouse. Even though farmers have to spend Rs one to two lakhs to cover the 25 percent cost of establishing one greenhouse, the greenhouse is in high demand. Therefore, the widespread use of Ladakh Greenhouse technology demonstrated the receptiveness of the Ladakhi farmers to adopting new technologies.

14.4 Management of yellow tail moth

The incidence of insect pests and diseases is low in Ladakh. However, a severe infestation of the Yellow Tail moth (*Euproctis similis*) was reported in the Dah-Hanu belt of the Leh district on apricot trees from 2013–2016. Complete defoliation of apricot trees was observed. Controlling the insect through an integrated approach, including the use of pesticides, was a must. Since large-scale use of chemical pesticides has not been adopted in the region, there was a need to create awareness about the importance and safe use of chemical pesticides. Since the Buddhist population dominates the infested villages, the main concern before us in implementing the management strategies was non-cooperation by the farmers. Killing insects is considered a sin, and there is no prior record of large-scale killing of insects to protect crops. However, during implementation, we found that the farmers are ready to control insect pests if adequately educated. The previously unknown pest was managed within two years, thus saving Rs 2.3 crore per year. The pest has not been reported since 2019 and therefore checked for its further spread (Stobdan et al., 2019).

My personal experience in transferring new technologies to the farmers' fields and managing a new insect pest is a compelling illustration that farmers in Ladakh are receptive to adopting new technologies. However, the involvement of the concerned Agriculture and Allied UT department officers right from the initial field trial stage is a must. On-field demonstration on farmers' fields is more effective than on research farms. Effective dissemination of new technologies among farmers in far-flung villages is possible only if the new technology is made a part of the government schemes. Changing the traditional agricultural practices in fruit growing is a major challenge in the region. Despite many years of imparting training and field demonstrations, fruit growers in the region do not follow standard fruit-growing practices such as training, pruning, manuring, and hand harvesting.

REFERENCES

Angchok D, Srivastava RB (2013). Strengthening the transfer of technology among the high-altitude community of Ladakh. *Indian J Ext Educ.*, 49: 72–79

Angmo S (2019). Crop diversification and vegetable productivity enhancement using black polyethylene mulch in trans-Himalayan Ladakh, India. PhD thesis submitted to Bharathiar University.

Angmo S, Dolkar D, Dolkar P, Kumar B, Stobdan T (2019). Growing watermelon in high altitude trans-Himalayan Ladakh. *Natl Acad Sci Lett.*, 42: 379–382

Stobdan T, Deen M, Gupta V, Raghuvanshi MS (2019). Integrated management of yellow tail moth (*Euproctis similis*) on apricot in Leh Ladakh India. LAHDC Leh, Ladakh

Chapter-15

Synergistic Project Execution: Learnings from Ladakh

When resources are scarce, people tend to develop mastery in judicious use of the available resources. People in Ladakh have developed a system of sharing limited human resources to cope with the harsh climatic condition and the limited resources. *Langde* is the system of collective workforce sharing so that agricultural activities can be completed without hiring laborers. *Rarzee* is the traditional system of community-based grazing of sheep and goats (*Ra* = goat), while *Barzee* is the community-based grazing of cattle (*Ba* = cow). Few individuals take responsibility for grazing the entire flock of the village. Irrigation is regulated by a *Churpon* (*Chu* = water; *Spon* = Lord), who is either appointed or elected from within the village. The *Churpon* ensures that the scarce water resource is carefully distributed so everyone can have their fair share of the precious resource to irrigate their lands.

The lessons from the traditional systems are of relevance even in the modern days. Ladakh has limited trained human resources in the agriculture sector to work on the diverse challenges. Adopting the 'copy and paste' model of technology intervention, where models adopted for lowlanders are applied to the Ladakh region, is ineffective (Angchok and Srivastava, 2013). There are a handful of agricultural scientists, and conducting research in diverse fields has become extremely difficult. However, during my two decades of working on solving emerging problems and introducing new technologies, I have witnessed working together of people's representatives, administrators, scientists, and extension workers as a team in the region. Some of the teamwork I have witnessed in recent years are mentioned below.

15.1 Management of a new insect pest

The incidence of insect pests and diseases is low in Ladakh. However, a severe infestation of the Yellow Tail moth (*Euproctis similis*) was reported in the Dah-Hanu belt of the Leh district on apricot trees during 2013–2016. The pest was new to Ladakh, and there is no report of the pest infesting fruit crops worldwide. Complete defoliation of apricot trees was observed. Monetary loss due to defoliating caterpillars in affected seven villages alone in Leh district was estimated to be Rs 223.5 lakhs in 2016. The insect pest is new to India and needs to be managed quickly to check its spread to other parts of the country. However, not much is known about the insect and its management.

Because of the monetary loss to the farmers and health hazards of *Euproctis similis*, a district-level committee was constituted by the Ladakh Autonomous Hill Development Council (LAHDC) Leh on 19 October 2016 to suggest control measures. Accordingly, the committee I headed brought out a detailed report and submitted it on 01 November 2016. Hon'ble Member of Parliament

Ladakh sanctioned Rs 25 lakhs, and preparation for pest management began the same month. Given the severe pest infestation in distantly located remote villages, no single agency could manage the pest. Therefore, all agencies working in agriculture and allied fields were consulted. State Departments such as Horticulture, Agriculture, and Forests; Leh-based research institutes such as DRDO-DIHAR, ICAR-CAZRI, and SKUAST-K; and non-governmental organizations such as LEHO, LeDEG, and TATA Trust volunteered to launch management drive in the affected villages. Each agency took full responsibility, from organizing awareness camps to managing the insect pests in the allotted villages. The research institutes designed the scientific management strategies, while the Horticulture Department took on the responsibility of arranging the inputs such as pesticides, spray pumps, metal clippers, etc. The same were procured and distributed to the above agencies for implementation. Since the apricot growers in the infested villages are primarily Buddhist, the main concern before us in implementing the management strategies was non-cooperation by the farmers. Killing insects is considered a sin, and there is no prior record of large-scale killing of insects to protect crops.

Community engagement was one of the most critical strategies for successfully managing the pest. Village-level discussions involving the elected representatives, village heads, and representatives from technical agencies proved fruitful in community mobilization. Notably, greater affiliation towards the elected representative and village heads resulted in a unanimous decision at the village level in the first meeting for concerted actions of the entire community in pest control. Since large-scale use of chemical pesticides has not been adopted in the region, there was a need to create awareness about the importance and safe use of chemical pesticides, particularly in highly infested villages. We trained volunteer youth from every village on safely using pesticides and power sprayers. The previously unknown pest was managed within two years, thus saving Rs 2.3 crore per year. The pest has not been reported since 2019 and therefore checked for its further spread (Stobdan et al., 2019). Therefore, successful pest management involving an integrated action of several governmental and non-governmental agencies serves as a model for managing major insect-pest and diseases in the region in the future.

Community engagement for management of insect-pest in Hanu belt

15.2 The organic farming movement

In 2019, the Ladakh Autonomous Hill Development Council (LAHDC), Leh, took its first step to embrace organic farming. In January 2019, a Study Group comprising elected representatives, officers from the Agriculture, Horticulture, Animal Husbandry, Sheep Husbandry, Cooperative, Soil Conservation, Command Area Development, and Forest Departments of LAHDC Leh, and scientists from Leh-based research institutes such as DRDO-DIHAR, SKUAST-K, and ICAR-CAZRI was constituted by the LAHDC Leh. Elected representatives, scientists, and officers worked as a team in preparing the vision document 'Mission Organic Development Initiative of Ladakh: Policy, Strategy, and Action Plan'. Based on the recommendation of the Study Group, the LAHDC Leh unanimously declared to become a fully organic district by 2025. Later, the LAHDC Leh constituted a committee comprising representatives from DRDO-DIHAR, Agriculture Department, Horticulture Department, Animal Husbandry Department, Sheep Husbandry Department, Cooperative Department, and the Krishi Vigyan Kendra to prepare the organic project proposal.

The key to the success of organic farming lies in farmers' support and strong political will. A survey showed that most farmers (86.3 percent) support the decision of LAHDC Leh to certify Leh as a fully organic District by 2025. Only 6.6 percent are against the decision, while 7.1 percent of the respondents are indecisive. The organic movement in Ladakh is progressing well with the strong support of the political leaders, the administration,

Agriculture and allied departments, research institutes, and the farming community. Working together as a team, the project 'Mission Organic Development Initiative of Ladakh' is currently a flagship program of UT Ladakh. All the agencies and departments are working together to achieve the common goal of certifying Ladakh as organic.

15.3 The Ladakh Greenhouse

Given the necessity of growing vegetables in winter, many passive solar greenhouses have been established in Ladakh since the 1980s. However, the traditional passive solar greenhouses in use have several limitations, and the need for an improved farmer-friendly passive solar greenhouse was felt. In January 2020, the Administration of Union Territory of Ladakh constituted a committee comprising officers from the Agriculture and Horticulture Departments of Leh and Kargil districts; and scientists from DRDO-DIHAR and SKUAST-K to recommend a greenhouse design that is most suitable for the Ladakh region. After brainstorming, the committee comprising members from various departments and agencies recommended the DRDO-DIHAR developed Ladakh Greenhouse as the most effective passive solar greenhouse for the Ladakh region. Later, a project entitled 'Improving Food Security and Promoting Livelihood through Ladakh Greenhouse' was initiated by UT Ladakh in March 2020. The Agriculture Department of UT Ladakh is currently establishing Ladakh Greenhouse on farmers' fields in Leh and Kargil districts. The teamwork of the administration, scientists at DRDO-DIHAR, and the officers of the Agriculture Department resulted in the effective transfer of the technology we developed at the research farm to the farmers' field in the shortest possible time. The greenhouse benefits farmers by improving food security, promoting livelihood, water and energy conservation, sustainable organic farming, and women empowerment.

15.4 Introduction of a new cash crop

My colleagues and I at DRDO-DIHAR developed agro-techniques for growing watermelon, a warm-season crop, in open fields in Ladakh. After the experimental trials at the research farm, it needs to be demonstrated on the farmers' field. We conducted the first demonstration on farmers' fields in collaboration with the Agriculture Department in Phey village in 2016. LEHO and LeGEG, Ladakh-based NGOs, also joined us in our field trial on farmers' fields in seven villages. The trials were a big success, and farmers could replicate the results obtained at our research farm. Field trials are time-consuming, and without the teamwork of various governmental and non-governmental agencies, farmers in the region may have to wait a few more years to benefit from our research work. Upon completing a successful two-year trial in the farmers' fields by various agencies, DRDO-DIHAR formally transferred the technology to the Agriculture Department, LAHDC Leh, on 06 October 2017. Watermelon is now grown commercially under open field conditions. Farmers earn Rs 10-12 lakh per hectare, approximately ten times more than cereal crops. Watermelon is currently being produced in 55 villages in the Leh district.

Besides the above examples, there are several other success stories of teamwork in Ladakh. GI tagging of

Raktsey Karpo apricots, the introduction of mulching technology for doubling vegetable productivity, and extending fruit crop planting season, are just to name a few. The traditional and modern days teamwork which the people of Ladakh have shown are lessons of relevance far beyond Ladakh itself.

REFERENCES

Stobdan T, Deen M, Gupta V, Raghuvanshi MS (2019). Integrated management of yellow tail moth (*Euproctis similis*) on apricot in Leh Ladakh India. LAHDC Leh, Ladakh

Angchok D, Srivastava RB (2013). Strengthening the transfer of technology among the high-altitude community of Ladakh. *Indian J Ext Educ.*, 49: 72–79.

Chapter-16

Ladakh: A Research Hub for Plant Scientists

The high mountain region of Ladakh is characterized by a rugged topography at an average altitude of over 3000 m asl. The region is characterized by extreme temperature variations, low precipitation, high wind velocity, sparse plant density, a thin atmosphere with high UV radiation, and a fragile ecosystem. Leh town's hottest month is July (25.8°C), and January is the coldest (-13.0°C). The maximum and minimum temperature during the cropping season (May-September) is 22.3°C and 9.4°C, respectively. The region lies in the rain shadow of the Himalayas and thus receives little rainfall. The region experiences intense sunlight and high UV-B radiation, with an average light intensity at noon during cropping season being 150 kilo lux. The UV-B radiation in Leh is double compared to Delhi (Singh et al., 2005). Therefore, the region is a unique place for conducting high altitude related research studies.

Despite being a far-flung area in the trans-Himalaya, Leh is well connected to Delhi, Chandigarh, Jammu, Srinagar, and several other cities in India by air. The highways leading to Leh are open between May and November. Leh City is located 420 km from Srinagar and 430 km from

Manali. Therefore, Ladakh is easily approachable by air as well as by road for conducting research studies.

The DRDO-Defence Institute of High Altitude Research (DIHAR) is the only research institute with its headquarters in Leh. Besides, there are regional stations of Sher-e-Kashmir University of Agricultural Sciences and Technology of Kashmir, ICAR-Central Arid Zone Research Institute, and Govind Ballabh Pant National Institute of Himalayan Environment. The University of Ladakh was also established in the year 2019.

Some of the areas, but not limited to, which researchers may find interesting to conduct studies include:

16.1 Plant response to altitudinal gradients

Altitudinal gradients are among the most powerful 'natural environments' for testing the effect of environmental factors on plant physiology and quality characteristics. Steep changes in temperature, moisture, atmospheric pressure, ultraviolet radiation, hours of sunshine, wind, season length, and geology occur along an altitudinal gradient (Körner, 2007). Accordingly, to better understand the effect of climatic variables, Ladakh, with an altitudinal gradient from 2400 to 5900 m, may serve as an excellent place for the researchers. We found that mulberry (*M. alba*) showed high phenotypic variation along an altitudinal gradient. A small change in altitude leads to a significant change in phenotype (Bajpai et al., 2015).

16.2 Plant adaptation to climate change

The current interest in plant adjustment to the environment results from an urgency to predict species responses to global climate change. Altitudinal gradients offer a means for testing biota's ecological and evolutionary responses to climate change. Understanding the role of phenotypic plasticity is necessary for predicting how plants will respond to climate change. Phenotypic plasticity in mulberry in the trans-Himalaya may serve as a model system to study plant adaptation to future environmental changes (Bajpai et al., 2014).

16.3 Cold hardiness

The ability of plants to adapt to and survive cold and freezing temperatures has many facets, which are often species-specific. It is the result of the response to many environmental cues rather than just low temperature. However, most of the cold tolerance responses in plants are studied in controlled conditions in greenhouses and growth chambers. Varying diurnal temperatures that produce mixed signals in the field are in contrast to the constant temperatures in controlled conditions. The research designed to explore and understand cold hardiness should be verified in the context of the physiology, growth habit, and life cycle of the plant grown under natural conditions. Ladakh presents an excellent opportunity to investigate the cold hardiness in natural conditions.

16.4 Plant response to hypoxia

Plants are aerobic organisms that require oxygen for their respiration. Hypoxia arises due to the insufficient availability of oxygen. Human adaptation and health-related issues in high mountain regions have been extensively studied. However, little is known about plant response to hypoxia in high-altitude regions. Ladakh thus offers an opportunity to conduct studies in this less-explored research field.

16.5 Combined tolerance to abiotic stress

Exposure to one type of stress is known to activate plant responses that facilitate tolerance to several different types of stress. Multiple environmental stresses are known to occur simultaneously in Ladakh. Therefore, researchers interested in studying combined tolerance to abiotic stresses may find the place befitting for their studies.

16.6 Plant-associated microbes in plant stress mitigation

Various environmental stresses cause adverse effects on plant growth, development, and productivity. In recent years, the role of plant-associated microbes in enhancing stress resistance and coping with the negative stress impact through the induction of various mechanisms has been established. Due to the stressful environmental conditions of Ladakh, the existence of novel plant-associated microbes and their role in enhancing stress resistance is worth exploring.

16.7 Plant physiology

The climatic condition of Ladakh offers exciting areas for researchers to study plant physiology under unique climatic conditions. High ultraviolet and photosynthetic radiation, low relative humidity, long day length, and high day and night temperature variation open doors for exciting research fields.

16.8 Nutrient uptake and assimilations

Researchers may find studies on plant nutrient uptake and assimilation interesting in Ladakh. Temperature is known to be critical to plant physiology and nutrient uptake. We observed that by applying 25 tonnes of farm yard manure per hectare, the yield of a selected crop is above the national average.

16.9 Flower color development

The flowers in Ladakh develop an intense color, an important determinant that directly influences their commercial value. The prevailing low temperature, photosynthetic irradiance, and high UV-B cause intense flower color (Ben-Tal and King, 1997). Water deficiency, common in Ladakh, also causes flowers to turn darker (Lai et al., 2011). Crop scientists can study various aspects of flower color development, including the expression of the gene of interest.

16.10 Fruit color development

The visual appearance of the fruit determines the perceived fruit quality. In general, better-colored fruits are in higher demand. We observed that fruits naturally formed intense skin color under the climatic conditions of Ladakh. The development of intense colored skin in the region may be due to the low temperature at the maturity stage. In general, low night temperatures and moderate daily temperatures promote red color development. The intensity and quality of sunlight is an important environmental factor that affects skin color. Light exposure, particularly UV-B radiation, induces the expression of anthocyanin biosynthesis regulating genes. Researchers interested in fruit color development may find Ladakh an appropriate study place.

16.11 Regulation of fruit sweetness

Genetic components influencing sugar contents and sugar profiles are well documented. However, little is known about the influence of environmental factors on fruit sweetness. We found that the geographical elevation markedly influenced fruit sugar contents in apricot in Ladakh. A linear relationship was observed between increasing altitude with total sugar, sucrose, and sorbitol contents (Naryal et al., 2019). Therefore, researchers keen to understand the regulation of fruit sweetness may find Ladakh the right place for conducting such studies.

16.12 Environmental contribution to pungency

Capsaicinoids are alkaloids causing the pungency in hot pepper fruit. This trait is a valuable fruit attribute for the food, pharmaceutical, and pesticide industries. When grown in Ladakh, we observed that the hot peppers have low pungency. It could be due to the dry climate or a combination of several abiotic and biotic factors. Researchers interested in studying environmental contribution to pungency may find Ladakh an exciting place for their studies.

16.13 Regulation of plant secondary metabolites

Plant secondary metabolites are known to play a major role in adapting plants to their changing environments but also represent an important source of active pharmaceuticals. Abiotic stress signals are known to influence secondary metabolites in plants. Researchers may find Ladakh an interesting site to study the time- and space-bound regulation of plant secondary metabolites.

16.14 The geographical barrier in gene flow

Geographic isolation is a major factor that restricts gene flow via pollen and seed dispersal. We found that the high mountain range is an effective barrier to gene flow in Ladakh (Bajpai et al., 2014). Researchers interested in studying the role of geographical obstacles in shaping the genetic structure of a plant species will find Ladakh an excellent place for such studies.

16.15 Natural barrier on the spread of insect pests and diseases

Understanding the patterns and trends in the spread of crop pests and pathogens worldwide is an important topic. The majority of the villages in Ladakh are separated from each other by high mountains. Researchers working on the distribution of crop pests and pathogens may find Ladakh an interesting site to study the role of the high mountain barrier on the spread of insect pests and diseases.

16.16 Doner genetic material for abiotic stress

The landraces found in Ladakh can be valuable resources for crop breeding, given their long selection history under stressful conditions. Landraces could be donors of important agronomically important traits, such as drought and cold tolerance. They represent significantly broader genetic diversity than modern varieties; therefore, they could contribute to extending the genetic base of the modern cultivars.

16.17 Breeding for multiple abiotic stress

The impact of climate change, drought, flooding, and other extreme events pose challenges to global crop production. Understanding the crop's ability to survive and maintain yield is important when faced with simultaneous exposures to multiple abiotic stresses. Since low and high temperatures, drought, high sun intensity and ultraviolet radiation, and oxidative stresses are natural components of Ladakh's ecosystem, the region offers an excellent opportunity to study plants' responses to multiple abiotic stresses .

16.18 Shuttle breeding

Shuttle breeding is an off-season field-testing technique whereby genetic material is grown in contrasting environments to turn over two plant generations per year. By implementing this simple yet effective technique, breeders have reduced the time required to complete a breeding cycle by 50 percent. Various cool-season crops that grow in winter on the mainland can be grown in summer in Ladakh. The Energy Resources Institute (TERI) and Delhi University have explored Ladakh for the shuttle breeding of mustard.

16.19 Multilocation trials

Multilocation trials play an important role in plant breeding and agronomic research. Breeders compare different crop genotypes to identify the superior ones, while agronomists use multilocation trials to compare combinations of agronomic factors. Ladakh's unique topography and climatic conditions present the scope for conducting multilocation trials.

16.20 Seed production

The Ladakh region is ideal for organic seed production due to its low humidity, bright sunshine, and low incidence of disease and insect pest infestation. True potato seeds (TPS) and garlic aerial bulbils are naturally formed in Ladakh. Researchers looking to produce high-quality seeds will find Ladakh a suitable place.

REFERENCES

Bajpai PK, Warghat AR, Sharma AR, et al. (2014) Structure and genetic diversity of natural populations of *Morus alba* L. in the trans-

Himalaya Ladakh region. *Biochem Genet.*, 52: 137–152.

Bajpai PK, Warghat AR, Yadav A, Kant A, Srivastava RB, Stobdan T (2015). High phenotypic variation in *Morus alba* L. along an altitudinal gradient in the Indian trans-Himalaya. *J Mt Sci.*, 12(2): 446–455

Ben-Tal Y, King RW (1997) Environmental factors involved in colouration of flowers of Kangaroo Paw. *Sci Hortic.*, 72: 35–48

Körner C (2007). The use of 'altitude' in ecological research. *Trends Ecol. Evol.*, 22: 569–574.

Lai YS, Yamagishi M, Suzuki T (2011) Elevated temperature inhibits anthocyanin biosynthesis in the tepals of an Oriental hybrid lily via the suppression of LhMYB12 transcription. *Sci Hortic.*, 132: 59–65

Naryal A, Acharya S, Bhardwaj AK, Kant A, Chaurasia OP, Stobdan T (2019). Altitudinal effect on sugar contents and sugar profiles in dried apricot (*Prunus armeniaca* L.) fruit. *J Food Compos Anal.*, 76: 27–32

Singh R, Nath S, Tanwar R, Singh S (2005). Study of erythemal dose variation and exposure time for different UV-B dose levels at Indian mainland and Antarctica. *Proc. URSI*, New Delhi

About The Author

Dr. Tsering Stobdan is a Ladakh-based agriculture scientist at the Defence Institute of High Altitude Research. He received formal education from pre-primary to secondary school at the Tibetan Children Village, Ladakh. He later completed his Bachelor's degree in Horticulture from the University of Horticulture and Forestry, Solan; Master's degree from GB Pant University of Agricultural Science and Technology, Pantnagar; and Ph.D. in Molecular Biology and Biotechnology from the Indian Agricultural Research Institute, New Delhi.

Dr. Stobdan has been at the forefront of providing appropriate agro-technologies to the farmers in Ladakh. The technologies he developed have reached over 120 villages in Ladakh. He identified native apricots (Raktsey Karpo) as the world's unique and sweetest apricot that later became Ladakh's first GI tag product. Based on his report, Seabuckthorn is included as an activity under the Mission for Integrated Development of Horticulture (MIDH) scheme of GoI. He introduced watermelon and sun melon as a new cash crop in Ladakh.

He has published over 80 research articles on topics related to horticulture and protected cultivation in Ladakh. Ten Research Fellows have been awarded Ph.D. degrees under his supervision.

Dr. Stobdan is a fellow of the Indian Academy of Horticultural Sciences (IAHS) and the Indian Society of Horticultural Research and Development. He is a recipient of the Fakhruddin Ali Ahmed Award for Outstanding Research in Tribal Farming Systems from the Indian Council of Agricultural Research. The Ladakh Greenhouse technology that he developed received the Prime Minister Award for Excellence in Public Administration (under the Innovation category) on 21 April 2022.

Printed in the USA
CPSIA information can be obtained
at www.ICGtesting.com
CBHW022006190824
13414CB00027B/157